# Basic Engineering
# and Strength of Materials

**MADAN MOHAN DAS**

*Formerly*
Professor, Civil Engineering Department
Assam Engineering College, Guwahati
Emeritus Fellow of AICTE
Director of Technical Education, Government of Assam

**MIMI DAS SAIKIA**

Associate Professor
Civil Engineering Department
Royal School of Engineering and Technology, Guwahati

**BHARGAB MOHAN DAS**

Chief Business Development Officer and Chairperson
Green Building Division
Ritta Company Ltd., Bangkok

**PHI Learning** Private Limited

New Delhi-110001
2010

₹ 225.00

**BASIC ENGINEERING MECHANICS AND STRENGTH OF MATERIALS**
Madan Mohan Das, Mimi Das Saikia, and Bhargab Mohan Das

ISBN-978-81-203-4181-4

The export rights of this book are vested solely with the publisher.

Published by Asoke K. Ghosh, PHI Learning Private Limited, M-97, Connaught Circus, New Delhi-110001 and Printed by Mudrak, 30-A, Patparganj, Delhi-110091.

# Contents

## Part II    STRENGTH OF MATERIALS

# Preface

The aim of this book is to present the course on **Basic Engineering Mechanics and Strength of Materials** covering the elementary topics of engineering mechanics and strength of materials introduced in the universities and the AICTE approved engineering and technological institutions of India for second semester students of all disciplines.

**Part I** (Chapters 1–11) of the book focuses on basic engineering mechanics. **Part II** (Chapters 12–17) deals with elementary knowledge of strength of materials.

Part I (Basic Engineering Mechanics) thoroughly explains system of forces, laws of mechanics and resultant forces, moments of forces, parallel forces and couples, equilibrium of forces, forces in space-introduction of vector algebra, analysis of forces in frames, centre of gravity, moment of inertia, friction and kinetics of rigid body.

In Part II (Strength of Materials), a detailed analysis of simple and generalized stress and strain, bending moment and shear force in beams, stresses in thin cylinders and shells, torsion and Euler's theory on column is given.

The contents of the book are presented in a concise form and in lucid language, keeping in view the course contents of the second semester students so that the book may prove useful to them. Students, in their first year of engineering career, while studying the above course, have to face a lot of difficulties in absence of a book that carries both the parts in a single volume and thus they face an uphill task to grasp the knowledge needed to start the basics of the course. To minimize this problem of budding engineering students of all branches, this book has been written in an easy-to-follow language to make it student-friendly and, of course, teacher-friendly. Equations are derived stepwise for beginners; numerical examples are solved in each and every chapter keeping in view its practical use in the field of engineering. Chapter-end problems help students sharpen their problem-solving skills.

We express our sincere thanks and appreciation to Mrs. Lalita Das, and Prof. L.P. Saikia of Royal School of Engineering and Technology, Guwahati, who have been a great source of encouragement to us. We would like to acknowledge our appreciation to Professor Mrinal Kumar Borah of Assam Engineering College for his help in the preparation of the manuscript. Our sincere thanks go to our other well-wishers who encouraged us to write this book.

Further, we would like to offer our profound respect to the late Professor Emeritus D.I.H. Barr of University of Strathclyde, Glasgow, who was our guide and philosopher since long.

We invite suggestions and criticisms from the readers for improving the contents of the book in subsequent reprints or editions.

**Madan Mohan Das**
**Mimi Das Saikia**
**Bhargab Mohan Das**

# Part I

## Basic Engineering Mechanics

# CHAPTER 1

# Introduction and System of Forces

## 1.1 INTRODUCTION

*Mechanics* is a branch of science which deals with the effect of forces on bodies and fluids. The body on which the force acts may be rigid or deformable. When mechanics is applied to different fields of engineering, it is called *engineering mechanics*. When forces act on a rigid body, there may be two possibilities. The body may remain static or may be in motion. Thus mechanics of rigid body is classified into statics and dynamics. *Statics* is that branch of mechanics which deals with study and behaviour of rigid bodies in the state of rest under the action of force while *dynamics* is another part in which the bodies are in the state of motion when the forces act on it. Again dynamics is further divided into kinematics and kinetics. In *kinematics*, the forces causing the motion are not considered while in *kinetics*, the forces causing the motion are considered. The mechanics of deformable bodies deals with the study of the bodies that undergo deformation under the action of forces. It may further be classified into strength of material, theory of plasticity and elasticity. *Fluid mechanics* is another branch of engineering mechanics. Substances like liquid, gas and vapour are fluids which have no definite shapes of their own but conform to the shape of the container. These fluids are capable of moving. Fluid continues to deform when subjected to shear force. Fluid mechanics is the study of these fluids at rest and in motion. This study deals with static, dynamic, kinematic and dynamic aspect of fluid. Study also takes account for conservation of mass, momentum, Newton's law of motion and laws of thermodynamics to make the mechanics of fluid interdisciplinary. Under the action of force, fluid may be compressible or incompressible. In mechanics, *force* plays the major role and, therefore, in this introductory chapter, discussion of different systems of forces has been presented in detail.

## 1.2 FORCES

The forces that act on bodies are thrusts or tensions, attraction or repulsion, friction, etc. Force is a *vector quantity*, hence it has direction and magnitude which is measured in newton in SI unit. In the field of mechanics, force is the most important factor. It is an agent which produces or tends to produce, destroys or tends to destroy the motion of body or particle. Different forms of force are as follows:

**Tension:**   As shown in Figure 1.1(a), a string is attached to a fixed end and on the other end a heavy load is suspended. The fibres of the string undergo a certain pull throughout its length which undergoes by the name *tension*. If the tension force is increased beyond a certain limit, it causes the string to break.

**Weight:**   Weight of body is the force with which earth attracts the body towards its centre. The direction of this force is vertical as shown in Figure 1.1(b).

**Reaction:**   In Figure 1.1(c), a load $W$ rests on a beam which has its support at $A$ and $B$. Due to this application of load, two forces, i.e. reactions $R_A$ and $R_B$ are being developed as shown in Figure 1.1(c).

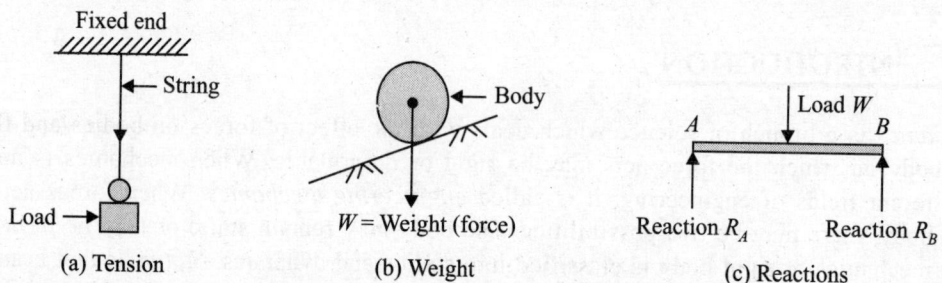

**Figure 1.1**   Different forms of force.

## 1.3   CHARACTERISTICS OF FORCES

A force has four characteristics as mentioned below.

   (i) Direction
   (ii) Magnitude
   (iii) Point $P$ on which it acts
   (iv) Line of action

All the above four characteristics are shown in Figure 1.2.

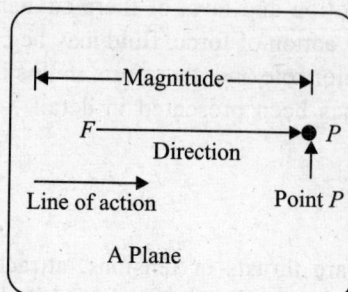

**Figure 1.2**   Characteristics of a force.

## 1.4    SYSTEM OF FORCES

When two or more forces act on a body, they are called *system of forces*. The following are different systems of forces.

   (i)  Coplanar force system
  (ii)  Non-coplanar force system
 (iii)  Collinear and non-collinear force system

### 1.4.1    System of Forces: Coplanar

**Coplanar forces:**  The forces, whose lines of action lie in the same plane are called **coplanar forces** (Figure 1.3(a)). Coplanar forces again may be coplanar collinear, coplanar concurrent, coplanar parallel, coplanar non-concurrent non-parallel forces.

**Coplanar concurrent forces:**  If the forces lie in the same plane and meet at the same point but they act in different directions, they are called **coplanar concurrent forces**. In Figure 1.3(b), the forces $F_1$, $F_2$ and $F_3$ are coplanar concurrent forces.

**Coplanar parallel forces:**  In Figure 1.3(c), the forces $F_1$, $F_2$ and $F_3$ lie in the same plane and act parallel to one another and they are called **coplanar parallel forces**.

**Coplanar non-concurrent non-parallel forces:**  In Figure 1.3(d) all the forces $F_1$, $F_2$, $F_3$, $F_4$, $F_5$ and $F_6$ are in the same plane. The lines of action are in the same plane but they are neither parallel nor meet at a same point. Hence, they are called **coplanar non-concurrent non-parallel forces**.

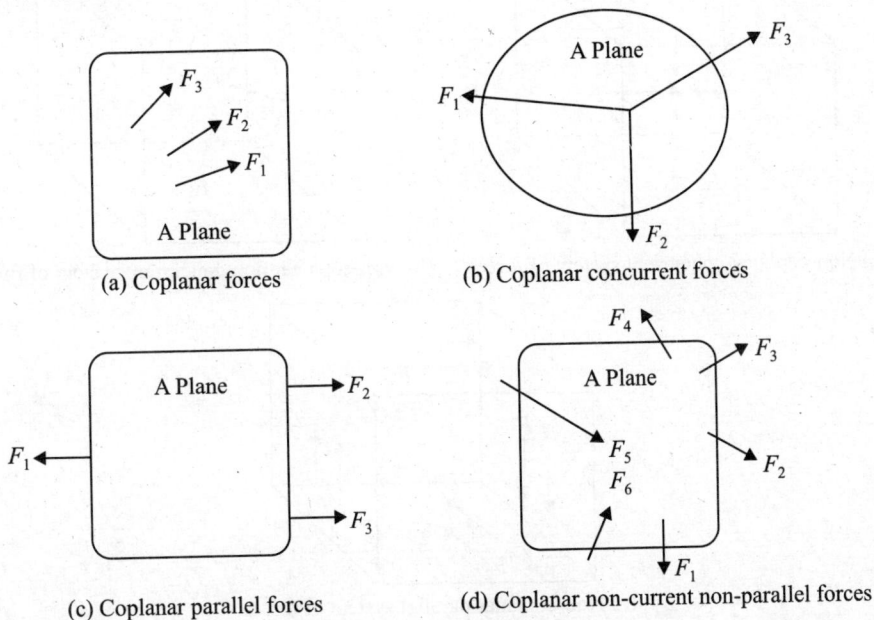

(a) Coplanar forces

(b) Coplanar concurrent forces

(c) Coplanar parallel forces

(d) Coplanar non-current non-parallel forces

**Figure 1.3**  Coplanar forces.

Again coplanar parallel system of forces may be like parallel and unlike parallel forces.

**Coplanar like parallel forces:**    These forces are shown in Figure 1.4(a). They act parallel in the same direction.

**Coplanar unlike parallel forces:**    They are represented in Figure 1.4(b). As shown in figure, $F_1$ and $F_2$ move parallel in the same directions, whereas $F_3$ and $F_4$ move parallel in both directions. All the four forces are parallel.

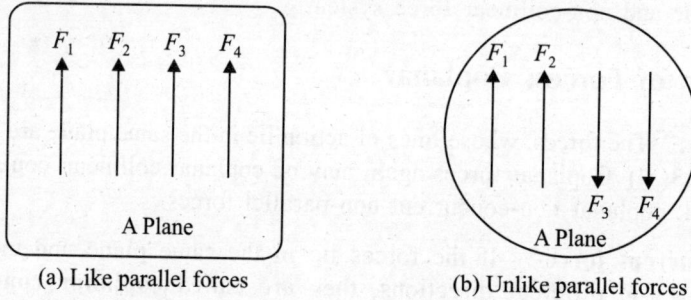

(a) Like parallel forces                    (b) Unlike parallel forces

**Figure 1.4**  Like and unlike parallel forces.

## 1.4.2   System of Forces: Non-coplanar

If the forces act in different planes, they are called **non-coplanar force system**. They may be further categorized as concurrent, non-concurrent and parallel system of forces. All these sub-divisions of non-planar forces are shown in Figure 1.5.

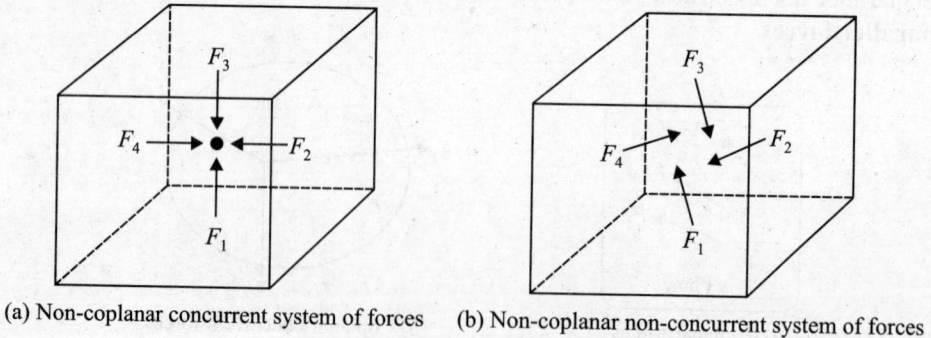

(a) Non-coplanar concurrent system of forces        (b) Non-coplanar non-concurrent system of forces

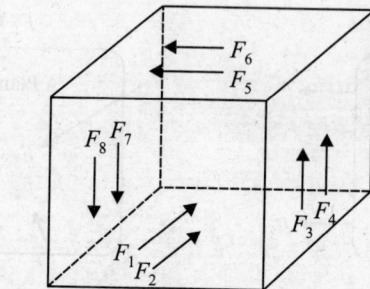

(c) Non-coplanar parallel system of forces

**Figure 1.5**  Non-coplanar system of forces.

### 1.4.3    System of Forces: Collinear and Non-collinear

When two or more forces coincide with one another, it is called a **collinear system of forces** (Figure 1.6). On the other hand, if the lines of action of the forces do not coincide, it is called **non-collinear system of forces** (Figure 1.7).

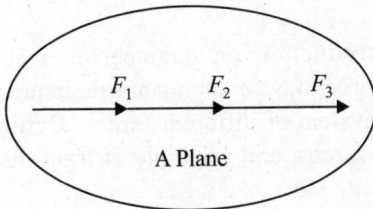

**Figure 1.6**   Collinear force system.        **Figure 1.7**   Non-collinear force system.

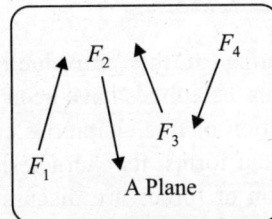

### 1.4.4    Principle of Transmissibility of Forces

This principle states that a force can be transmitted from one point to another point along the same line of action such that the effect produced by the force on the body remains unchanged.

In Figure 1.8, a rigid body is subjected by a force $F$ at $C$. According to the principle of transmissibility, the same force $F$ can be transmitted to a new point $C'$ along the same line of action such that net effect remains unchanged.

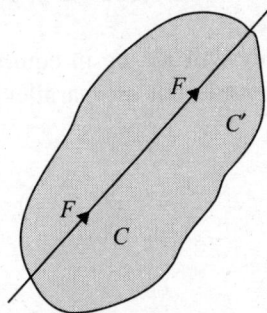

**Figure 1.8**   Transmissibility of forces.

## 1.5   PRINCIPLES OF EQUILIBRIUM OF FORCES

The following three common principles of equilibrium of forces are worth to be mentioned.

1. **Two-force principle:**   It states that if a body in equilibrium is acted upon by two forces, then they must be equal, opposite and collinear.
2. **Three-force principle:**   It states that if a body in equilibrium is acted upon by three forces, then resultant of any two forces must be equal, opposite and collinear with the third force.

**3. Four-force principle:**  It states that if a body in equilibrium is acted upon by four forces, then resultant of any two forces must be equal, opposite and collinear with the resultant of the other two.

## 1.6   CONCLUSION

In the beginning of this introductory chapter, introduction on engineering mechanics, its different sub-disciplines have been defined. Since force plays a dominant role in mechanics, main presentation of this chapter is concentrated on system of different forces. Definition of force in different forms, its characteristics, system of forces, and principle of transmissibility and equilibrium of forces are discussed.

## EXERCISES

1.1  Define mechanics and explain its various sub-disciplines.

1.2  Explain tension, weight and reactions with simple sketches.

1.3  Draw a sketch to show the characteristics of forces.

1.4  Enumerate the different systems of coplanar forces with sketches.

1.5  Explain the system of non-coplanar forces with figures.

1.6  Explain the principle of transmissibility and equilibrium of forces.

1.7  Two forces are acting on a body and the body is in equilibrium. What conditions should be satisfied by these two forces?

1.8  How will you prove that a body will not be in equilibrium when it is subjected to two forces which are equal and opposite but are parallel?

CHAPTER $2$

# Laws of Mechanics and Resultant Forces

## 2.1 INTRODUCTION

Different laws of mechanics starting from Newton's laws, gravitational law, and all other laws to determine the resultant forces are presented in this chapter. All these laws are powerful tools to solve most of the engineering and applied sciences problems with which engineers and scientists have to encounter. Applications of different laws are demonstrated with the help of quite a number of numerical problems in engineering fields.

## 2.2 LAWS OF MECHANICS

The basic laws of mechanics and their principles used in mechanics are as follows:

(i) **Newton's first law:** states that every body continues in a state of rest or uniform motion in a straight line unless it is compelled to change that state by some external force acting on it.

(ii) **Newton's second law of motion:** states that the net external force acting on a body in a direction is directly proportional to the rate of change of momentum in that direction.

(iii) **Newton's third law:** states that every action has always an equal and opposite reaction.

(iv) **Gravitational law of attraction:** which states that two bodies will be attracted to each other along their connecting line with force which is directly proportional to the product of their masses and inversely proportional to the square of the distance between their centres. If $m_1$ and $m_2$ are the masses of the two bodies, $r$ and $F$ are the distance and force of attraction between them, then according to this law:

$$F \propto m_1 \cdot m_2$$

and
$$F \propto \frac{1}{r^2}$$

9

or

$$F \propto \frac{m_1 m_2}{r^2}$$

∴

$$F = G\frac{m_1 m_2}{r^2}$$

where $G$ is the universal gravitational constant of proportionality and its value is $6.67 \times 10^{-11}$ Nm²/kg².

   (v) **Law of parallelogram forces:**   This law is explained with figures in Section 2.4 in detail.

  (vi) **Law of triangle of forces:**   This one is also explained in Section 2.5.

 (vii) **Law of polygon of forces:**   This law is presented in Section 2.6.

## 2.3   RESULTANT FORCE

If a body is acted upon by different forces $M$, $N$, $P$, and $Q$ simultaneously, it is possible to find out one single force $R$ which could replace them, i.e. this single force can produce the same effect as produced by all the forces. This single force $R$ is the resultant force of $M$, $N$, $P$ and $Q$.

## 2.4   LAW OF PARALLELOGRAM FORCES

It is the law to find the resultant of two concurrent coplanar forces. It states that *if two forces acting simultaneously on a particle, be represented in magnitude and direction by adjacent sides of a parallelogram, their resultant may be represented in magnitude and direction by the diagonal of the parallelogram, which passes through their point of intersection.*

   Consider two forces $P$ and $Q$, acting at $O$, represented by the straight line $OA$ and $OC$ in magnitude and direction as shown in Figure 2.1. If a parallelogram $OABC$ is drawn with $OA$ and $OC$ as adjacent sides, the resultant $R$ of these two forces may be represented by the diagonal $OB$. Let $\alpha$ be the angle of the two forces $P$ and $Q$. Produce $OA$ and drop perpendicular $BD$ on this extended line in the same Figure 2.1. From the geometry of the figure, it is known that:

$$AB = Q \quad \text{and} \quad \angle BAD = \theta$$
$$BD = Q \sin\theta \quad \text{and} \quad AD = Q \cos\theta$$

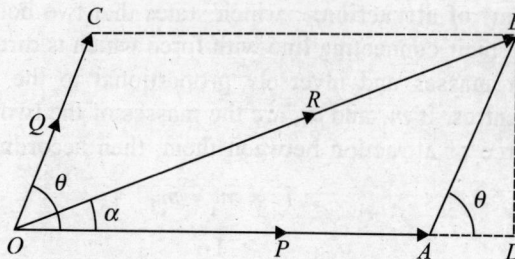

**Figure 2.1**   Parallelogram law of forces.

In the right-angled triangle $OBD$:

$$OB = R = \sqrt{OD^2 + BD^2}$$

$$= \sqrt{(OA + AD)^2 + BD^2}$$

$$\therefore \quad R = \sqrt{(P + Q\cos\theta)^2 + (Q\sin\theta)^2}$$

$$R = \sqrt{P^2 + Q^2\cos^2\theta + 2PQ\cos\theta + Q^2\sin^2\theta}$$

$$R = \sqrt{P^2 + Q^2(\sin^2\theta + \cos^2\theta) + 2PQ\cos\theta}$$

$$R = \sqrt{P^2 + Q^2 + 2PQ\cos\theta} \tag{2.1}$$

If the resultant $OB$ makes angle $\alpha$ with $OD$, then

$$\tan\alpha = \frac{BD}{OA + AD} = \frac{Q\sin\theta}{P + Q\cos\theta}$$

$$\tan\alpha = \frac{Q\sin\theta}{P + Q\cos\theta} \tag{2.2a}$$

$$\therefore \quad \alpha = \tan^{-1}\left(\frac{Q\sin\theta}{P + Q\cos\theta}\right) \tag{2.2b}$$

***EXAMPLE 2.1***   Two forces act at an angle of 100°. The bigger force is 50 N and the resultant is perpendicular to the smaller one. Find this smaller force.

***Solution***   Considering Figure 2.2,

$\theta = 100°$, $P = 50$ N, the angle between the resultant $R$ and smaller force $Q$ is 90°

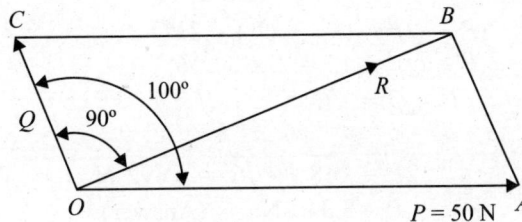

**Figure 2.2**   Visual of Example 2.1.

Hence angle between the 50 N force and the resultant $= (100° - 90°) = 10°$
Assuming $Q$ to be smaller and using Eq. (2.2), we have

$$\tan\alpha = \frac{Q\sin\theta}{P + Q\cos\theta}$$

$$\tan 10° = \frac{Q\sin 100°}{50 + Q\cos 100°} = \frac{Q\times 0.9848}{50 - 0.173648\,Q}$$

or
$$0.176327 = \frac{Q \times 0.9848}{50 \quad 0.173648\,Q}$$

or
$$8.81635 - 0.0306\,Q = 0.9648\,Q$$

or
$$0.9954\,Q = 8.81635$$

∴
$$Q = 8.857 \text{ N} \qquad \text{(Answer)}$$

**EXAMPLE 2.2**   The resultant of two forces is $\sqrt{12}$ N when the forces act at right angles. The resultant of the same forces is $\sqrt{15}$ N if they act at angle of 60°. Find the magnitudes of the two forces.

**Solution**   Let the forces be $P$ and $Q$ and the resultant is $R$.
Considering Figure 2.1, if $\theta = 90°$, $R = \sqrt{12}$ N and if $\theta = 60°$, $R = \sqrt{15}$ N
Using Eq. (2.1),

$$R = \sqrt{P^2 \quad Q^2 \quad 2PQ\cos\theta}$$

or
$$\sqrt{12}\text{ N} = \sqrt{P^2 \quad Q^2 \quad 2PQ\cos 90°}$$

or
$$12 = (P^2 + Q^2) \qquad \because \cos 90° = 0 \tag{i}$$

Similarly,

$$\sqrt{15} = \sqrt{P^2 \quad Q^2 \quad 2PQ\cos 60°}$$

or
$$15 = (P^2 + Q^2) + 2PQ \times 0.5$$

or
$$15 = 12 + PQ \qquad \because (P^2 + Q^2) = 12 \text{ by (i)}$$

or
$$PQ = 3 \tag{ii}$$

We know that
$$(P + Q)^2 = (P^2 + Q^2) + 2PQ$$

or
$$(P + Q)^2 = 12 + 2 \times 3 = 18$$

or
$$P + Q = \sqrt{18} \tag{iii}$$

Similarly

$$(P - Q)^2 = (P^2 + Q^2) - 2PQ$$

or
$$(P - Q)^2 = 12 - 2 \times 3 = 6$$

$$P - Q = \sqrt{6} \tag{iv}$$

Solving (iii) and (iv),

$$2P = \sqrt{18} + \sqrt{6} = 6.692 \text{ N}$$

∴
$$P = 3.346 \text{ N} \qquad \text{(Answer)}$$

and
$$Q = \sqrt{18} - 3.346 = 0.8966 \text{ N}$$

∴
$$Q = 0.8966 \text{ N} \qquad \text{(Answer)}$$

**EXAMPLE 2.3**   Find the magnitude of two equal forces acting at a point with an angle of 60° between them, the resultant $50 \times \sqrt{3}$ N.

**Solution**   Using Eq. (2.1),

$$R = \sqrt{P^2 \quad Q^2 \quad 2PQ\cos\theta}$$

Here $P = Q$ as the two forces are equal and $\theta = 60°$. Therefore, Eq. (2.1) becomes,

$$R = \sqrt{P^2 + P^2 + 2P^2 \cos 60°}$$

$$R = \sqrt{2P^2(1 + \cos 60°)}$$

or

$$50 \times \sqrt{3} = P\sqrt{2}(1 + \cos 60°)$$

Solving for $P = 48.825$ N
Hence two equal forces are $P = Q = 48.825$ N    (Answer)

**EXAMPLE 2.4**    Two forces acting at a point have their resultant 10 N when they act at right angle and their least resultant is 2 N. Find their greatest resultant and also the resultant when they act at an angle of $\theta = 60°$.

**Solution**    Let $P$ be the greater than force $Q$. When they act at $\theta = 90°$, the resultant $R$ becomes:

$$R = 10 = \sqrt{P^2 + Q^2}$$

or

$$P^2 + Q^2 = 100$$

Also their least resultant is:

$$2 = (P - Q)$$

$$P^2 + Q^2 - 2PQ = 4$$

$$100 - 2PQ = 4$$

$$\therefore \qquad PQ = 96/2 = 48$$

Now the greatest resultant is:

$$R_{max} = P + Q = \sqrt{P^2 + Q^2 + 2PQ}$$

$$= \sqrt{100 + 96} = 14 \text{ N} \qquad \text{(Answer)}$$

When they act at $\theta = 60°$,

$$R = \sqrt{P^2 + Q^2 + 2PQ \cos 60°}$$

$$= \sqrt{100 + 2 \times 48 \times 0.5}$$

$$= \sqrt{148} \text{ N}$$

$$= 12.165 \text{ N} \qquad \text{(Answer)}$$

## 2.5    LAW OF TRIANGLE OF FORCES

The two forces $P$ and $Q$ are acting simultaneously on a particle (Figure 2.3(a)) and they are represented both in magnitude by the two sides of a triangle taken in order, the magnitude and direction of the resultant $R$ can be represented by the third sides of the triangle $AC$ in opposite direction as shown in Figure 2.3(b).

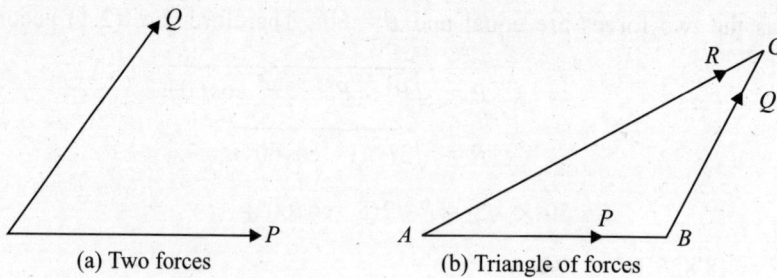

(a) Two forces  (b) Triangle of forces

**Figure 2.3** Law of triangle of forces.

## 2.6 LAW OF POLYGON OF FORCES

Polygon law of forces is the extension of law of triangle of forces. It states that if a number of forces acting simultaneously on a particle, be represented in magnitude and direction, by the sides of a polygon taken in order, then the resultant of all these forces may be represented in magnitude and direction, by the closing side of the polygon, taken in opposite order.

In Figure 2.4(a), five forces $A$, $B$, $C$, $D$ and $E$ are shown with their directions with arrow heads. In Figure 2.4(b), these five forces $A$, $B$, $C$, $D$ and $E$ are drawn in magnitude and directions with the names $a$, $b$, $c$, $d$ and $e$ respectively. The closing side $ST$ of the polygon $PQRST$ taken in opposite order with its direction shown by arrow head represents the resultant.

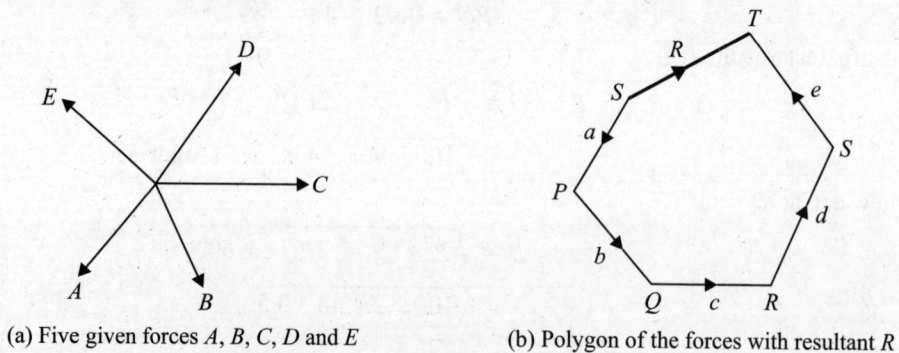

(a) Five given forces $A$, $B$, $C$, $D$ and $E$  (b) Polygon of the forces with resultant $R$

**Figure 2.4** Details of polygon of forces.

## 2.7 RESOLUTION OF A FORCE: COMPONENTS OF A FORCE

Resolution of forces means to determine the components of given force in two directions in the same plane. In Figure 2.5, a given force $R$ makes an angle of $\theta$ with $X$-axis. Resolving this force along $X$-axis and $Y$-axis, the components are $R \cos\theta$ and $R \sin\theta$ respectively. Thus resolution of forces is the process of finding the components in specified directions of co-ordinate system.

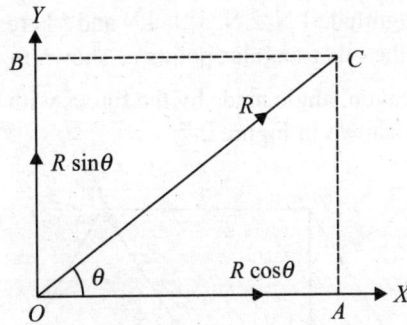

**Figure 2.5** Resolution of a force.

## 2.8  RESULTANT OF A NUMBER OF COPLANAR FORCES

A number of coplanar forces $F_1$, $F_2$, $F_3$, ... are acting at point $O$ at angles $\theta_1$, $\theta_2$, $\theta_3$, ... respectively with $X$-axis as shown in Figure 2.6.

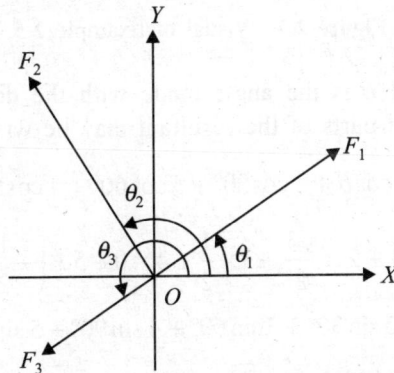

**Figure 2.6**  Resultant of a number of coplanar forces.

Let $H$ be the summation of horizontal components of the forces along $X$-axis.

Let $V$ be the summation of the vertical components of the forces along $Y$-axis.

Let $R$ be the resultant of all the forces acting at an angle of $\theta$ with the $X$-axis.

All the forces $F_1$, $F_2$, $F_3$, ... are resolved into two components in $X$-axis and $Y$-axis as shown in Section 2.8

then $\qquad\qquad\qquad \Sigma H = F_1 \cos\theta_1 + F_2 \cos\theta_2 + F_3 \cos\theta_3 \ ...$ (2.3 a)

and $\qquad\qquad\qquad \Sigma V = F_1 \sin\theta_1 + F_2 \sin\theta_2 + F_3 \sin\theta_3 \ ...$ (2.3 b)

Now resultant of the forces:

$$R \quad \sqrt{H^2 \quad V^2} \qquad\qquad (2.4)$$

The angle made by resultant $R$ with the $X$-axis $\theta$ is given by:

$$\tan \quad \frac{\Sigma V}{\Sigma H} \qquad\qquad (2.5)$$

**EXAMPLE 2.5**   Forces of magnitude 1 N, 2 N, 3 N, 4 N and 5 N respectively act at angular points of a regular hexagon towards the other angular points as shown in Figure 2.7. Find the resultant.

**Solution**   Since a regular hexagon, angle made by the forces with direction of 1 N force goes on increasing by 30° each time as shown in Figure 2.7.

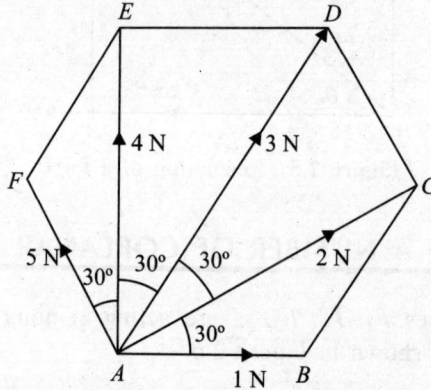

**Figure 2.7**   Visual of Example 2.5.

Let $R$ is the resultant and $\theta$ is the angle made with the direction 1 N force, and then algebraic sum of the resolved parts of the resultant may be written:

$$R \cos\theta = 1 + \cos\theta + 2\cos 30° + 3\cos 60° + 4\cos 90° + 5\cos 120°$$

$$= 1 + 2 \times \frac{\sqrt{3}}{2} + 3 \times \frac{1}{2} + 4 \times 0 + 5 \times \left(-\frac{1}{2}\right) = \sqrt{3}\ \text{N}$$

$$R \sin\theta = 2\sin 30° + 3\sin 60° + 4\sin 90° + 5\sin 120°$$

$$= 2 \times \frac{1}{2} + 3 \times \frac{\sqrt{3}}{2} + 4 \times 1 + 5 \times \frac{\sqrt{3}}{2} = 5 + 4\sqrt{3}\ \text{N}$$

$\therefore$
$$R^2 = \left(\sqrt{3}\right)^2 + \left(5 + 4\sqrt{3}\right)^2 = 76 + 40\sqrt{3} = 4\left(19 + 10\sqrt{3}\right)$$

or
$$R = 2\sqrt{19 + 10\sqrt{3}} = 12.053\ \text{N}  \qquad \text{(Answer)}$$

and
$$\tan\theta = \frac{5 + 4\sqrt{3}}{\sqrt{3}} = 6.886$$

$$\theta = 81.738°  \qquad \text{(Answer)}$$

**EXAMPLE 2.6**   The two forces are acting on a bolt at $A$ as shown in Figure 2.8. Determine the magnitude of the resultant.

**Solution**   Resolve the two forces $P$ and $Q$ from Figure 2.8 in horizontal and vertical directions as:

$$\Sigma H = P\cos 25° + Q\cos 135° = 65\cos 25° + 90\cos 135° = -5.4367\ \text{N}$$

and
$$\Sigma V = P\sin 25° + Q\sin 135° = 65\sin 25° + 90\sin 135° = 91.1098\ \text{N}$$

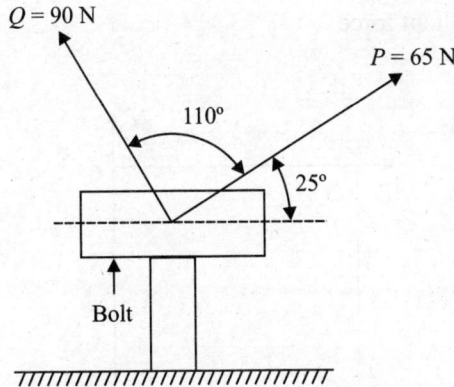

**Figure 2.8**   Visual of Example 2.6.

Now resultant of the forces:

$$R = \sqrt{\Sigma H^2 + \Sigma V^2} = \sqrt{(-5.4367)^2 + (91.1098)^2} = 91.2718 \text{ N} \qquad \text{(Answer)}$$

***EXAMPLE 2.7***   Three forces equal to $3P$, $7P$ and $5P$ act simultaneously along the three sides of an equilateral triangle $ABC$ of side $a$. Find the magnitude and direction of the resultant force.

***Solution***   Let $R$ be the resultant force (Figure 2.9)

Resolving the forces horizontally,

$$\Sigma H = 7P \sin30^\circ + 5P \sin30^\circ - 3P$$
$$= 3.5P + 2.5P - 3P$$
$$= 3P$$

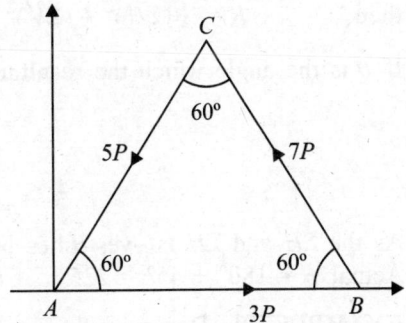

Resolving the forces vertically,

$$\Sigma V = 7P \cos30^\circ + 5P \cos30^\circ$$
$$= 7P\frac{\sqrt{3}}{2} - 5P\frac{\sqrt{3}}{2}$$

**Figure 2.9**   Visual of Example 2.7.

$$= 2\frac{\sqrt{3}}{2}P = \sqrt{3}P$$

$$R = \sqrt{(\Sigma H)^2 + (\Sigma V)^2} = \sqrt{(3P)^2 + \left(\sqrt{3}P\right)^2} = \sqrt{12P^2} = \sqrt{4 \times 3}P = 2\sqrt{3}P \qquad \text{(Answer)}$$

If $\theta$ is the angle which the resultant makes with horizontal, then,

$$\tan\theta = \frac{\Sigma V}{\Sigma H} = \frac{\sqrt{3}P}{3P} = \frac{1}{\sqrt{3}}$$

$$\therefore \qquad\qquad \theta = 30^\circ \qquad \text{(Answer)}$$

***EXAMPLE 2.8***   Four forces equal to $P$, $2P$, $3P$ and $4P$ are acting along the four sides of a square $ABCD$ respectively taken in order as shown in Figure 2.13. Find the magnitude and direction of the resultant force.

*Solution*    Let $R$ be the resultant force
Consider Figure 2.10.

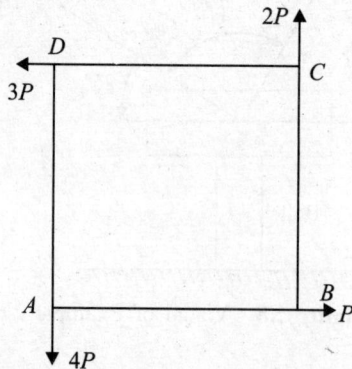

**Figure 2.10**    Visual of Example 2.8.

Resolving the forces horizontally;

$$\Sigma H = P - 3P = -2P$$

and

$$\Sigma V = 2P - 4P = -2P$$

then

$$R = \sqrt{(\Sigma H)^2 + (\Sigma V)^2} = \sqrt{(-2P)^2 + (-2P)^2} = 2\sqrt{2}P \qquad \text{(Answer)}$$

If $\theta$ is the angle which the resultant makes with horizontal, then,

$$\tan \theta = \frac{\Sigma V}{\Sigma H} = \frac{-2P}{-2P} = 1$$

$$\therefore \qquad \theta = 45^\circ$$

As the $\Sigma H$ and $\Sigma H$ is –ve, $\theta$ lies between $180^\circ$ and $270^\circ$,
Actual $\theta = 180^\circ + 45^\circ = 225^\circ$    (Answer)

**EXAMPLE 2.9**    Determine the resultant of the coplanar concurrent forces shown in Figure 2.11.

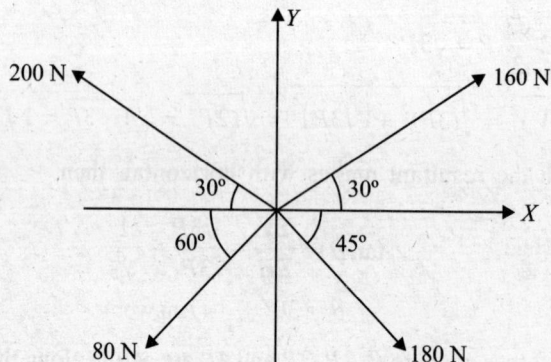

**Figure 2.11**    Visual of Example 2.9.

*Solution*    Using the Eqs. (2.3a) and (2.3b) of Section 2.8,

$\Sigma H$ is the summation of horizontal components of the forces.

Hence

$\Sigma H = 160 \cos 30° + 200 \cos 150° + 80 \cos 240° + 180 \cos 315°$

or    $\Sigma H = 138.56 - 173.20 - 40.0 + 127.28 = 52.64$ N

$\Sigma V$ is the summation of vertical components of the forces.

Hence

$\Sigma V = 160 \sin 30° + 200 \sin 150- + 80 \sin 240° + 180 \sin 315°$

or    $\Sigma V = 80 + 100.0 - 69.28 - 127.28 = -16.56$ N

Now resultant of the forces:

$$R = \sqrt{\Sigma H^2 + \Sigma V^2} = \sqrt{(-16.56)^2 + (52.64)^2} = 54.89 \text{ N} \qquad \text{(Answer)}$$

## 2.9    CONCLUSION

Various laws of mechanics are defined and explained in detail with figures. Equations are derived for different laws. Their applications are demonstrated with numerical examples for clear understanding of the readers. The laws or theorems discussed in this chapter have lots of applications in strength of materials, theory of structures and applied sciences.

## EXERCISES

**2.1** State and explain the following laws:
   (a) Laws of parallelogram of forces
   (b) Law of triangle of forces
**2.2** Explain with diagrams the coplanar collinear forces and coplanar parallel forces.
**2.3** State different laws of mechanics.
**2.4** Find the angle between two equal forces $P$ when their resultant is
   (a) equal to $P$ and
   (b) equal to $P/2$                                    (Answer   (a) 120° (b) 151°3′)
**2.5** Three forces of $2P$, $3P$, $4P$ act at a point in directions parallel to three sides of an equilateral triangle taken in order. Find the magnitude and line of action of the resultant force.                                    (Answer   $P\sqrt{3}$, 210)
**2.6** Two forces 15 N and 12 N are acting at a point. The angle between the forces is 60°. Find the value of the resultant.                                    (Answer   20.43 N)
**2.7** The system of forces is shown in Figure 2.12. The resultant is 240 N acting along $Y$-axis. Compute the magnitude of $P$ and its inclination with $X$-axis.

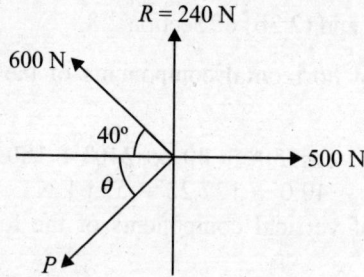

**Figure 2.12**    Visual of Exercise 2.7.

(Answer   $P = 151.16$ N, $\theta = 74.51°$)

**2.8** Two forces $P + Q$ and $P - Q$ make an angle of $2\alpha$ with one another and their resultant makes an angle $\theta$ with bisector of the angle between them. Show that,

$$P \tan\theta = Q \tan\alpha$$

**2.9** Two forces $P$ and $Q$ acting on a particle at an angle $\alpha$ have a resultant $(2k + 1)\sqrt{P^2 + Q^2}$. When they act an angle of $90° - \alpha$, the resultant becomes $(2k - 1)\sqrt{P^2 + Q^2}$. Prove that

$$\tan\alpha = \frac{k - 1}{k + 1}$$

**2.10** The three forces $P$, $Q$ and $R$ in one plane act on a particle. The angles between $Q$ and $R$, $R$ and $P$, $P$ and $Q$ are $\alpha$, $\beta$ and $\gamma$ respectively. Show that their resultant is equal to $(P^2 + Q^2 + R^2 + 2QR \cos\alpha + 2RP \cos\beta + 2PQ \cos\gamma)$.

# CHAPTER 3

# Moments of Forces

## 3.1 INTRODUCTION

Forces acting on a rigid body may produce either a motion of translation or rotation or translation and rotation both. The case of rotation introduces the idea of *turning effect* or *moment of a force*. The use or application of moment is a powerful tool to solve many practical problems occurring in the field of engineering and applied sciences.

## 3.2 MOMENT OF A FORCE

A moment of a force about a point is defined as the product of the force and the perpendicular distance of the point from the line of action of the force. To have a clear understanding of moment, consider the following definition sketch (Figure 3.1).

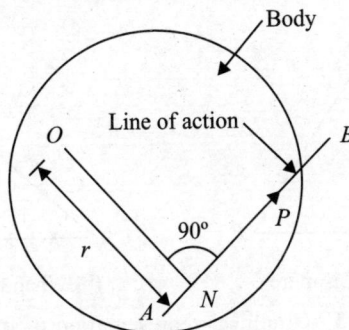

**Figure 3.1** Anticlockwise moment.

As shown in Figure 3.1, a force $P$ acts along the line of action $AB$ about the point $O$ of the body. Let $r$ be the perpendicular distance of line of action of the body, i.e. $ON$ is perpendicular to $AB$. Then moment $M$ of the force $P$ about the point $O$ of the body is given by

$$M = P \times r \tag{3.1}$$

The tendency of the moment is to rotate the body in anticlockwise direction about $O$. Hence, this moment is called *anticlockwise moment*.

If the force $P$ acts about the point $O$ of the body clockwise, as shown in Figure 3.2, it is called *clockwise moment*.

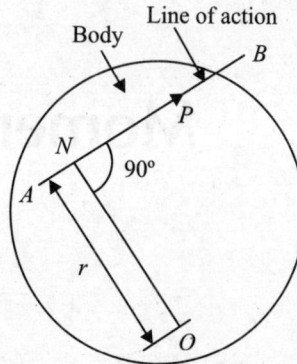

**Figure 3.2**   Clockwise moment.

The anticlockwise moment shown in Figure 3.1 is considered +ve, then clockwise moment shown in Figure 3.2 is taken –ve. If the force is in $N$ and $ON$ is in m, unit of moment is N = m.

## 3.3   GRAPHICAL OR GEOMETRICAL REPRESENTATION OF MOMENT

Let the force $P$ be represented in magnitude and direction along the line of action $AB$ both in Figures 3.3(a) and 3.3(b).

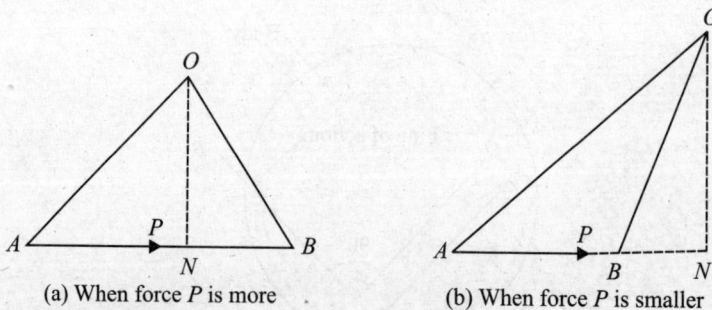

(a) When force $P$ is more          (b) When force $P$ is smaller

**Figure 3.3**   Graphical representation of moment.

Moment is taken about the point $O$. Drop perpendicular from $O$ to $AB$ or $AB$ produced. Join line $OA$ and $OB$. Now moment of the force $P$ about the point $O$ is $P \times ON = AB \times ON = 2\Delta OAB$. Thus the magnitude of the moment of a force about a point is represented by twice the area of the triangle formed by joining the point to the extremities of the line representing the force.

## 3.4   LAW OF MOMENTS

Law of moments states that if a number of forces, all being in one plane, are acting at a point in equilibrium, the sum of clockwise moments must be equal to the sum of anticlockwise moments. In other words, algebraic sum of the moments of forces about any point in equilibrium condition of the body is equal to zero. This is an important law in the field of statics and is used in different applied sciences and engineering fields.

## 3.5   VARIGNON'S THEOREM

It states that the algebraic sum of moments of two forces (which do not form a couple) about any point in their plane is equal to the moment of their resultant about that point.

**Proof**   Consider Figure 3.4 in which the point lies outside the two forces $P$ and $Q$ in Figures 3.4(a) and in 3.4(b), $O$ lies outside the two forces. In both the cases, the two forces $P$ and $Q$ act at point $A$ along $AX$ and $AY$ respectively and let $O$ be any point in their plane about which moments are taken.

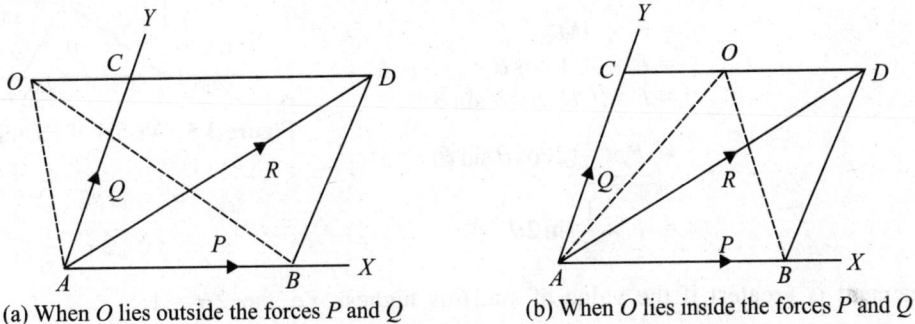

(a) When $O$ lies outside the forces $P$ and $Q$        (b) When $O$ lies inside the forces $P$ and $Q$

**Figure 3.4**   Diagram to prove Varignon's theorem.

Draw $OC$ parallel to $P$ to meet the line of action of $Q$ at $C$. Choose a suitable scale so that $AC$ represent the magnitude of $Q$ and with same scale, let $AB$ represent the magnitude of $P$. Compute the parallelogram $ABDC$. Join $AD$, $OA$, and $OD$. Now $AD$ represents the resultant $R$ of $P$ and $Q$. Now both in Figure 3.4(a) and Figure 3.4(b), moments of $P$, $Q$ and $R$ about $O$ are represented by $2\Delta OAB$, $2\Delta OAC$ and $2\Delta OAD$ respectively.

In Figure 3.4(a) where $O$ lies outside the $\angle BAC$, the moments of $P$ and $Q$ are both of the same sign and their algebraic sum is:

$$2\Delta OAB + 2\Delta OAC = 2\Delta DAB + 2\Delta OAC$$
$$= 2\Delta CAD + 2\Delta OAC$$
$$= 2\Delta OAD$$
$$= \text{Moment of } R \text{ about } O$$

In Figure 3.4(b) where $O$ lies within the $\angle BAC$, the moments of $P$ and $Q$ are of opposite sign. If we consider moment of $P$ +ve then moment of $Q$ will be −ve. Now their algebraic sum is:

$$2\Delta OAB - 2\Delta AOC = 2\Delta DAB - 2\Delta AOC = 2\Delta CAD - 2\Delta OAC$$
$$= 2\Delta OAD$$
$$= \text{Moment of } R \text{ about } O$$

Extension of Varignon's theorem has led to the generalized theorem of moments which states that if any number of coplanar forces acting on a rigid body have a resultant, this resultant is equal to the algebraic sum of the moments of those forces in their plane.

**EXAMPLE 3.1**  One end of stout rope of 4 m is fixed to a vertical telegraph post standing on the ground and a man pulls at the other end with given force $F$. Find the point of the post at which the rope is to be fixed in order that the man will have the best chance of overturning the post.

**Solution**  In Figure 3.5, $AB$ is the telegraph post, $CD$ is the rope by which the man is pulling with force $F$. $D$ is the position of the man on the ground.

$CD$ is given to be 4 m. From $A$ draw perpendicular $AM$ on $CD$.

Let $\angle ADC$ be $\theta$ and $F$ is the force applied by the man along $CD$.

Now moment of force $F$ about $A$

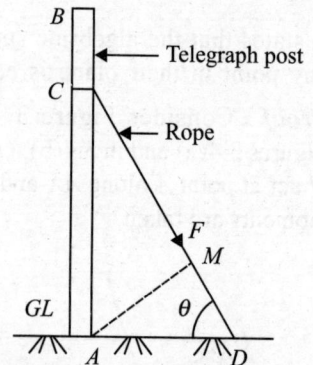

**Figure 3.5**  Visual of Example 3.1.

$$= F \times AM$$
$$= F \times AD \cos\theta$$
$$= F \times CD \cos\theta \sin\theta$$
$$= F \times \frac{1}{2}(2\cos\theta \sin\theta)$$
$$= F \times \frac{1}{2}\sin 2\theta$$

This moment is greatest if the value of $\sin 2\theta$ is highest, i.e. $\sin 2\theta = 1$

$$\therefore \qquad\qquad\qquad\qquad 2\theta = 90°$$
$$\therefore \qquad\qquad\qquad\qquad \theta = 45°$$
$$AC = CD \sin 45° = 4 \times \frac{1}{\sqrt{2}} = 2\sqrt{2} \text{ m} = 2.28284 \text{ m} \qquad \text{(Answer)}$$

**EXAMPLE 3.2**  The resultant $R$ of the water force $P$ and force $W$ due to weight of gravity dam is shown in Figure 3.6 with its magnitude, direction and distance from the toe of the dam. Find its moment about point $T$.

**Solution**  Resultant force at $C = 650$ kN

Draw perpendicular from toe $T$ on the line of action of force $CAB$ at $B$ as shown in Figure 3.6. Now triangle $TBC$ is a right-angled triangle and angle $TAB = 45°$

Now
$$\sin 45° = \frac{AB}{AT}$$

or
$$AB = AT \sin 45°$$

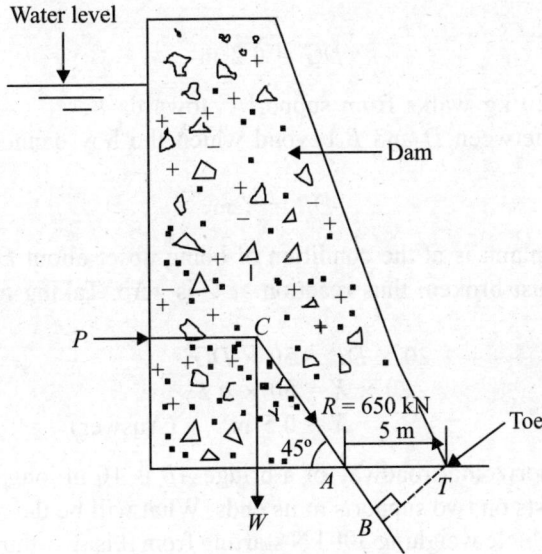

**Figure 3.6** Visual of Example 3.2.

$$= 5 \times 0.7071$$
$$= 3.3554 \text{ m}$$

Moment of 650 kN about $T = 850 \times 3.3554$
$$= 2852.09 \text{ kN-m} \qquad \text{(Answer)}$$

**EXAMPLE 3.3** A narrow uniform plank of length 2 m long weighs 50 kg. It is supported in horizontal position by two supports, one 0.5 m from one end and other 0.8 m from other end of the plank. A boy weighing 20 kg walks on the plank starting from the latter post towards the corresponding end. Find, how far it is safe for him to walk.

**Solution** As shown in Figure 3.7,

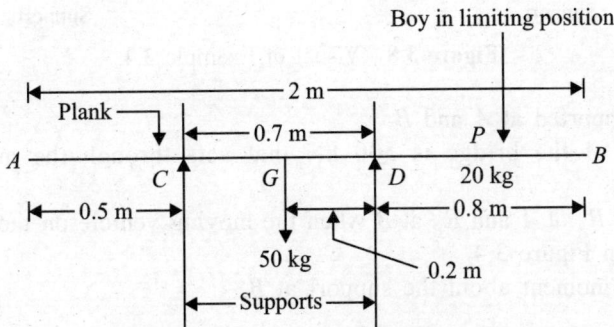

**Figure 3.7** Visual of Example 3.3.

Let $AB$ the plank placed on the supports $C$ and $D$. It is given that $AC$ is 0.5 m and $BD$ is 0.8 m. As the plank is 2 m long, the distance between the supports is 0.7 m.

The weight (50 kg) of the plank acts at the middle of the beam at $G$.

Hence

$$DG = 0.2 \text{ m}$$

The boy who weighs 20 kg walks from support $C$ towards $B$.

Let $P$ be the point between $D$ and $B$ beyond which the boy cannot walk safely.

Let

$$DP = X \text{ m}$$

In this position, the plank is at the condition of being upset about $D$ and the contact of the plank at support $C$ is just broken, thus reaction at $C$ is zero. Taking moment about $D$ at this instant,

$$20 \times DP = 50 \times DG$$
$$20 \times X = 50 \times 0.2$$
$$X = 0.5 \text{ m} \quad \text{(Answer)}$$

**EXAMPLE 3.4**   The horizontal roadway of a bridge $AB$ is 10 m long. The self-weight of the bridge is 500 kN and rests on two supports at its ends. What will be the reactions at the supports when a moving heavy vehicle weighing 300 kN starting from $A$ is two-thirds of the way across the bridge?

**Solution**   In Figure 3.8, $AB$ is the bridge which is 10 m long.

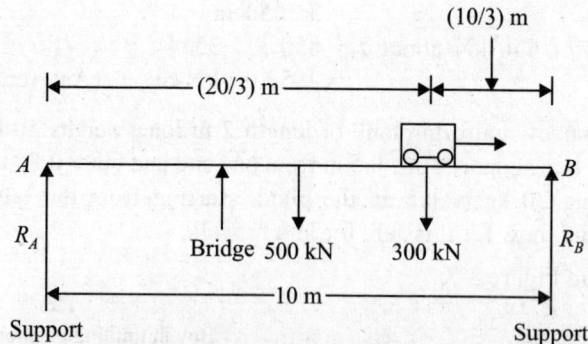

**Figure 3.8**   Visual of Example 3.4.

The bridge is supported at $A$ and $B$.

The self-weight of the bridge is 500 kN that acts through the middle as shown in Figure 3.8.

Find the reaction $R_A$ at $A$ and $R_B$ at $B$ when the moving vehicle on the bridge reaches the position as shown in Figure 3.4.

To find $R_A$, take moment about the support at $B$,

$$500 \times 5 + 300 \times (10/3) = R_A \times 10$$

or
$$R_A = \frac{500 \times 5 + 300 \times \dfrac{10}{3}}{10} = 350 \text{ kN} \quad \text{(Answer)}$$

Now

$$R_A + R_B = (500 + 300) \text{ kN} = 800 \text{ kN} \quad \text{(Answer)}$$

$$\therefore \quad R_B = (800 - R_A) = (800 - 350) \text{ kN} = 450 \text{ kN} \quad \text{(Answer)}$$

## 3.6   LEVERS

A lever is a rigid bar which is hinged at one end or points. It is free to rotate about this hinge end which is known as *fulcrum*. A lever has a point for effort called the *effort arm* and another point for lifting the load called the *load arm*.

### 3.6.1   Simple Lever

The lever which consists of one bar having one fulcrum is known as *simple lever* as shown in Figure 3.9. The lever may be straight, curved or even bent.

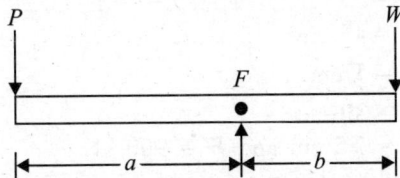

**Figure 3.9**   Simple lever showing fulcrum, effort arm and load arm.

In Figure 3.9, $P$ is the effort applied on effort arm of length $a$, and $W$ is the load applied in load arm of length $b$.

Equating the moment at about fulcrum $F$,

$$P \times a = W \times b$$

or

$$\frac{W}{P} = \frac{a}{b}$$

The terms $\dfrac{W}{P}$ and $\dfrac{a}{b}$ are commonly called as *mechanical advantage* and *leverage of the lever*.

In order to increase the mechanical advantage, either the length of $a$ is to be increased or the length of $b$ is to be decreased.

### 3.6.2   Compound Lever

Compound lever is the combination of a number of simple levers. It is shown in Figure 3.10. In compound lever, mechanical advantage or leverage is much greater than that of a simple lever.

Leverage of compound lever = Leverage of first lever + Leverage of the second lever + ⋯

An example of compound lever is the platform weighing machine. It is also used in godowns and parcel offices of transport agencies weighing the consignment of goods.

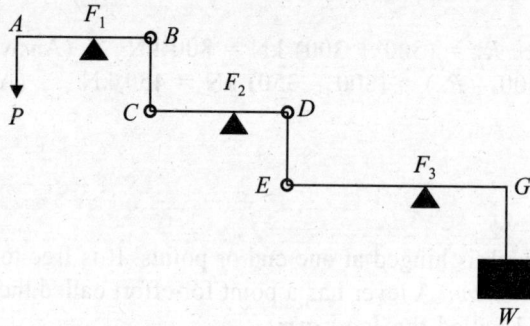

**Figure 3.10**   Compound lever.

***EXAMPLE 3.5***   A compound lever, shown in Figure 3.11, is required to lift a load of 900 kN. Find the effort at $A$ required to lift the load. Take $AF_1 = 45$ cm, $F_1B = 5$ cm, $CD = 30$ cm, $DF_2 = 7.5$ cm.

***Solution***   Given that:

$$AF_1 = 45 \text{ cm},$$
$$F_1B = 5 \text{ cm}$$
$$CD = 30 \text{ cm}$$
$$DF_2 = 7.5 \text{ cm and } W = 900 \text{ kN}$$

Let the effort at $A$ be $P$

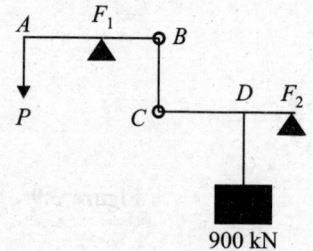

Now leverage of the upper lever     $AB = \dfrac{AF_1}{BF_1} = \dfrac{45}{5} = 9$

Similarly leverage of the lower lever $CF_2 = \dfrac{CF_2}{DF_2} = \dfrac{37.5}{7.5} = 5$

**Figure 3.11**   Visual of Example 3.5.

$\therefore$   Total leverage of the compound lever = $9 \times 5 = 45$

We have the relation

$$\frac{W}{P} = \text{Total leverage} = 45$$

or

$$\frac{900}{P} = 45$$

$\therefore$

$$P = \frac{900}{45} = 20 \text{ kN}$$

The effort at $A = P = 20$ kN     (Answer)

## 3.7   CONCLUSION

The chapter presents moment of a force, graphical or geometrical representation of moment, laws of moment, Varignon's theorem and its proof, both simple and compound lever where application of moments is required. Various numerical examples are solved to understand the theoretical presentation.

## EXERCISES

**3.1** What is meant by moment of a force? Explain it mathematically.

**3.2** How the moment of a force is represented geometrically?

**3.3** Explain with sketches the clockwise and anticlockwise moments.

**3.4** Enumerate the laws of moment.

**3.5** State and proof Varignon's theorem.

**3.6** Define lever and leverage of lever. Distinguish between simple and compound lever.

**3.7** $AB$ is the diameter of a circle. $AD$, $AC$ are two chords at right angles to each other. Show that the moments of the forces represented by $AC$, $AD$ about be are equal.

**3.8** A uniform beam $AB$ is 1.6 m long and weighs 50 N, weights of 20 $n$ and 50 $n$ are suspended from $A$ and $B$ respectively. At what point beam must be supported so that it may rest horizontally.      (Answer:   0.6 m from $B$)

**3.9** Prove that if four forces acting along the sides of a square are in equilibrium, they must be equal in magnitude.

**3.10** A heavy carriage wheel of weight $W$ and radius $r$ is to be dragged over an obstacle of height $h$, by a horizontal force $P$ applied in the centre of the wheel. Show that $P$ must

be slightly greater than $W \dfrac{\sqrt{2hr - h^2}}{r - h}$.

# CHAPTER 4

# Parallel Forces and Couples

## 4.1  INTRODUCTION

In Chapter 1, a brief discussion with figures on coplanar parallel forces, coplanar like parallel and unlike parallel forces, non-coplanar parallel forces has been presented. It is observed that parallel forces do not meet at one point although they have some effects on the body on which they act. Two equal and opposite parallel forces, i.e. equal and unlike forces acting on a body, do not have any resultant force. Thus it is concluded that no single force can replace two equal and opposite forces whose lines of action are different. Such a set of two equal and opposite forces, whose lines of action are different form a *couple*. A couple is thus unable to produce any translator motion in a straight line; but a couple produces rotation in the body on which it acts.

## 4.2  RESULTANT OF TWO LIKE UNEQUAL PARALLEL FORCES

Let two like unequal parallel forces act at points $A$ and $B$ respectively of a rigid body represented the lines $AX$ and $BY$ (Figure 4.1). Join $AB$.

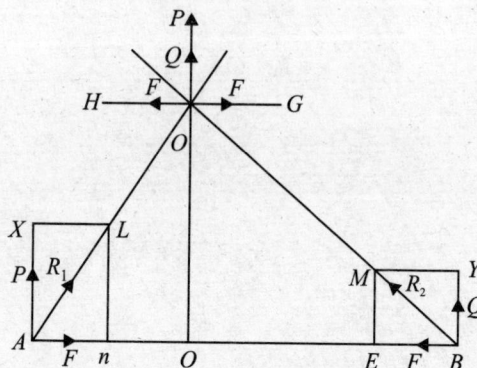

**Figure 4.1**  Resultant of two like unequal parallel forces.

30

At $A$ apply a force of any magnitude $F$ along $AB$ and at $B$ apply an equal and opposite force $F$ along $BA$ as shown in Figure 4.1. Since these two forces balance each other, they do not affect the required resultant. Let these two forces represent $AD$ and $BE$. Complete the parallelogram $ADLX$ and $BEMY$. The diagonals $TL$ and $BM$ are produced to meet at $O$. Through $O$, draw $OC$ parallel to $AX$ or $BY$ to meet $AB$ in $C$ and draw $HOG$ parallel to $AB$. Now the forces $P$ at $A$ and $Q$ at $B$ are equivalent to the forces $P$ and $F$ at $A$ and $Q$ and $F$ at $B$ since Force $F$ at $A$ and $B$ are equal and opposite having no effect in the system. But the forces $P$ and $F$ are equivalent to their resultant $R_1$ represented by their diagonal $AL$. Let the point of application this resultant $R_1$ be transferred to its line of action to $O$. $R_1$ at $O$ is resolved into two component forces parallel to their original directions, $F$ along $OG$, i.e. $AC$ and $P$ along $CO$ produced parallel to $AX$. Similarly Resultant $R_2$ of forces $Q$ and $F$ is resolved into two components at $O$, $F$ in the direction of $OH$, i.e. parallel to $CA$ and $Q$ along $CO$ produced parallel to $BY$. Thus the given forces are equivalent to forces $P$ and $Q$ along $CO$ and two more forces each equal to $F$ acting in opposite directions $OH$ and $OH$. The first two forces are equivalent to a single force $(P + Q)$ along $CO$ and last forces balance each other being equal and opposite.

Hence, the resultant $R$ of the two like parallel forces $P$ and $Q$ is a like parallel force $(P + Q)$ acting through a point $C$ in $AB$ between the points of application of $P$ and $Q$.

The position of the point $C$ through which $R$ acts may be obtained as:

Triangles $ACO$ and $ADL$ are similar

$\therefore$
$$\frac{AC}{CO} = \frac{AD}{DL} = \frac{AD}{AX} = \frac{F}{P} \tag{4.1}$$

Similarly from similar triangles $BDC$ and $BEM$,

$$\frac{BC}{CO} = \frac{BE}{EM} = \frac{BE}{BY} = \frac{F}{Q} \tag{4.2}$$

Dividing Eqs. (4.1) by (4.2),

$$\frac{AC}{BC} = \frac{Q}{P} \tag{4.3}$$

i.e. $C$ divides the line $AB$ internally in the inverse ratio of the forces $P$ and $Q$. If $P = Q$, $C$ is the midpoint of $AB$.

## 4.3  RESULTANT OF TWO UNLIKE UNEQUAL PARALLEL FORCES

The two unequal forces $P$ and $Q$ $(P > Q)$ are acting on a rigid body at $A$ and $B$ respectively. They are represented by the lines $AX$ and $AY$ in magnitude as shown in Figure 4.2. The two points $A$ and $B$ are joined. At $A$ and $B$ apply forces $F$ along $AB$ and $BA$. Since these two forces are equal and opposite, they will not affect the resultant. These two forces are represented by $AD$ and $BE$ respectively. Complete the parallelograms $ADLX$ and $BEMY$ just like the Figures 4.1 in Section 4.2. The diagonals of these two parallelograms meet at $O$ when produced. Draw line $OC$ parallel to $AX$ or $BY$. Line $BA$ when extended meets $OC$ at $C$.

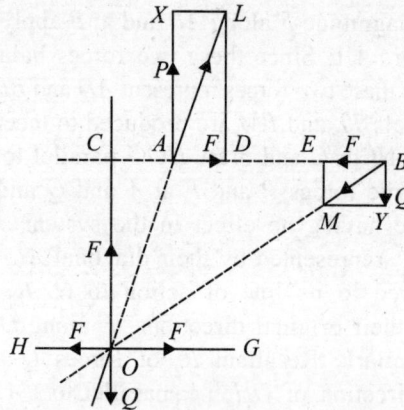

**Figure 4.2** Resultant of two unlike unequal parallel forces.

Proceeding similarly like the cases of resultant forces in Section 4.2, it can be shown at point $O$ that the resultant of these two unequal forces $P$ and $Q$ is equal to $(P - Q)$ as equal $F$ forces acting at $O$ opposite to each other balance each other. This resultant force $R = (P - Q)$ acts in the direction of greater force $P$ as shown in Figure 4.2 and this resultant meets at $C$ which lies outside the line $BA$, i.e. outside the points of application of $P$ and $Q$. The position of the $C$ where $R$ acts, can be obtained in the similar ways as used in Section 4.2 as:

Triangles $OCA$ and $AXL$ are similar

$\therefore$ 
$$\frac{OC}{AC} = \frac{AX}{XL} = \frac{AX}{AD} = \frac{P}{F}$$

$\therefore$ 
$$P \cdot AC = F \cdot OC \tag{4.4}$$

Similarly from similar triangles $MEB$ and $BEM$,

$$\frac{OC}{CB} = \frac{ME}{EB} = \frac{BY}{BE} = \frac{Q}{F}$$

$\therefore$ 
$$Q \cdot CB = F \cdot OC \tag{4.5}$$

Dividing Eq. (4.4) by Eq. (4.5),

$$\frac{AC}{OB} = \frac{Q}{P} \tag{4.6}$$

i.e. $C$ divides the line $AB$ externally in the inverse ratio of the forces $P$ and $Q$.

## 4.4 RESULTANT OF A SYSTEM OF PARALLEL FORCES

### Case I: When the forces are all like-parallel

Let $P_1$, $P_2$, $P_3$ ... be a system of like-parallel forces. To find the resultant of all this forces, first find out the resultant $R_1$ of the first two forces $P_1$, $P_2$. $R_1$ will be like-parallel force.

Then find the resultant $R_2$ of $R_1$ and $P_3$. $R_2$ will be like-parallel force. Then $R_2 = R_1 + P_3 = P_1 + P_2 + P_3$. $R_3$ is again a like-parallel force. This way final resultant of all the like-parallel forces will be:

$$R = P_1 + P_2 + P_3 + \cdots + P_n \qquad (4.7)$$

**Case II:   When the forces are not like-parallel**

Divide the forces into two sets of like-parallel forces.

Let $R_1$ be the resultant of the first set of like-parallel forces.

Let $R_2$ be the resultant of the second set like-parallel forces opposite to the first set. Obviously $R_1$ and $R_2$ will be unlike-parallel forces.

Now if $R_1 > R_2$, then $(R_1 - R_2)$ will be required resultant of all the forces which is parallel to $R_1$.

If $R_1 = R_2$, and their lines of action are coincident, the system is in equilibrium: but if their lines of action are not coincident, they form a *couple*.

**EXAMPLE 4.1**   A straight rod of 7 m long weighing 30 kN is carried by two men. One man supports it at a distance of 1.5 m from one end and the other at a distance of 2 m from the other end. What load each man does bear?

*Solution*   Let $AB$ be the 7 m rod. $G$ is the middle point so that the weight 50 kN acts through it. The supports of the first and second man are at $D$ and $C$ respectively at distance of 1.5 m and 2 m as shown in Figure 4.3.

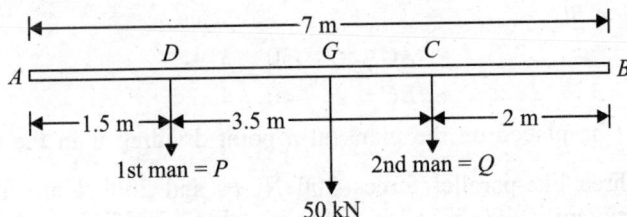

**Figure 4.3**   Visual of Example 4.1.

Take $P$ and $Q$ are the downward pressures on the first and second man at $D$ and $C$ respectively. Now $P$ kN and $Q$ kN are the parallel components of the weight 50 kN.

Now according to the resultant of like-parallel forces,

$$P + Q = 50 \text{ kN} \qquad (4.8)$$
$$P \times DG = Q \times CG$$
$$P \times (3.5 - 1.5) = Q \times (3.5 - 2)$$
$$2P = 1.5Q \qquad (4.9)$$

Putting the value of $P$ in Eq. (4.8), $0.75\,Q + Q = 50$

$$\therefore \qquad Q = \frac{50}{1.75}$$
$$= 28.571 \text{ kN} \qquad \text{(Answer)}$$
$$P = 50 - 28.571$$
$$= 21.429 \text{ kN} \qquad \text{(Answer)}$$

**EXAMPLE 4.2** A block of stone weighing 70 kN is being carried by two men on a plank. Where the block is to be placed so that one man has to carry more than 10 kN than the other man?

**Solution** Let $C$ be the point on the plank $AB$ where the stone is to be placed (Figure 4.4).

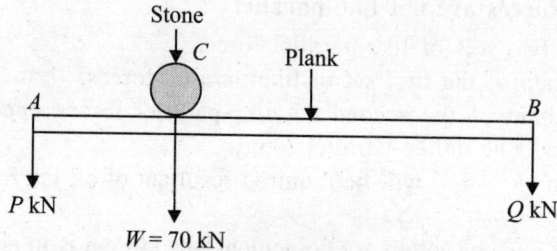

**Figure 4.4** Visual of Example 4.2.

Let $P$ and $Q$ are the loads ($P > Q$) which the two men at $A$ and at $B$ have to carry.
As seen from Figure 4.4, loads $P$, $W$ and $Q$ are unlike-parallel forces.
Now

$$P + Q = W = 70 \text{ kN}$$
$$P - Q = 10 \text{ kN}$$

Solving, $P = 40$ kN and $Q = 30$ kN
Again $P \times AC = Q \times BC$

or

$$\frac{AC}{BC} = \frac{Q}{P} = \frac{30}{40} = \frac{3}{4}$$

Thus the stone must be placed on the plank at a point dividing it in the ratio of 3:4

**EXAMPLE 4.3** Three like-parallel forces 100 N, $F$, and 300 N are acting as shown in Figure 4.5. If the resultant is 600 N and is acting at a distance of 45 cm from $A$, then find the magnitude of force $F$ and its distance from $A$.

**Solution** Considering Figure 4.5,

$$\text{Resultant } R = 100 \text{ N} + F + 300 \text{N}$$
$$600 \text{ N} = 100 \text{ N} + 300 \text{ N} + F$$
∴
$$F = 600 \text{ N} - 100 \text{ N} - 300 \text{ N} = 200 \text{ N} \quad \text{(Answer)}$$

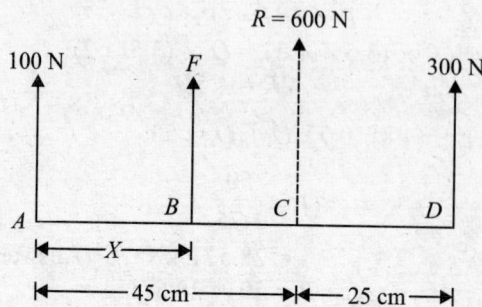

**Figure 4.5** Visual of Example 4.3.

To find the distance $X$ from $A$

Taking moment about $A$,

$$600 \times 45 = 200 \times X + 300 \times (45 + 25)$$

$$\therefore \qquad X = \frac{600 \times 45 - 300 \times 70}{200} = 30 \text{ cm} \qquad \text{(Answer)}$$

**EXAMPLE 4.4** Four parallel forces of magnitude 200 N, 200 N, 50 N, and 400 N are shown in Figure 4.6. Determine the magnitude of the resultant and also determine distance of the resultant from $A$.

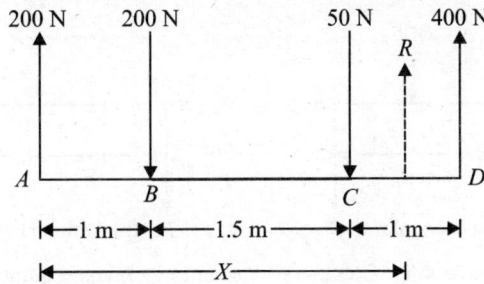

**Figure 4.6** Visual of Example 4.4.

**Solution** Let $R$ be the resultant of the unlike-parallel forces shown in Figure 4.6.
Then

$$R = 200 \text{ N} - 200 \text{ N} - 50 \text{ N} + 400 \text{ N} = 350 \text{ N} \qquad \text{(Answer)}$$

Let this resultant is at a distance $X$ from $A$

Taking moment about the point,

$$R \times X = -200 \times 1 - 50 \times 2.5 + 400 \times 3.5$$
$$350 \times X = -200 - 125 + 1400$$

$$\therefore \qquad X = \frac{-325 + 1400}{350} = \frac{1075}{350} = 3.071 \text{ m} \qquad \text{(Answer)}$$

## 4.5 COUPLE

Two equal and unlike-parallel forces cannot be represented by a single force. Moreover these two forces acting on a body, cannot produce a motion of translation of the body. Thus these two equal and unlike forces (whose lines of action are not the same) are said to constitute a *couple* as shown in Figure 4.7. The perpendicular distance $AB = d$ between the lines of action of two forces $P$ are called the *arm* of the couple. The *moment M* of the couple shown in

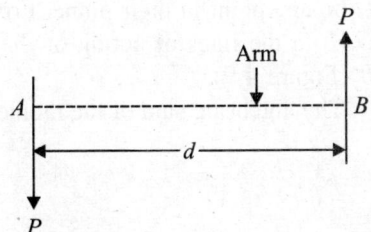

**Figure 4.7** Couple and its arm.

Figure 4.7 is the product of either of the forces forming the couple and the perpendicular distance between the forces, i.e. *AB* or *d*.

Mathematically,

$$M = P \times d$$

### 4.5.1   Classification of Couple

The couples are classified as clockwise and anticlockwise couples, depending on the direction in which they tend to rotate the body on which they act. These two types of couples are illustrated in Figure 4.8.

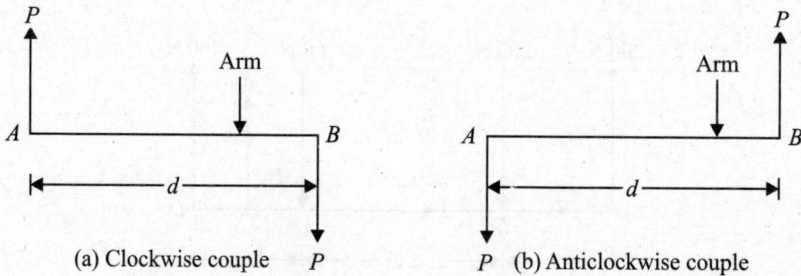

(a) Clockwise couple    *P*      *P* (b) Anticlockwise couple

**Figure 4.8**   Clockwise and anticlockwise couples.

### 4.5.2   Characteristics of Couple

Couple irrespective of clockwise or anticlockwise has some characteristics. They are as follows:

(i) Algebraic sum of the forces constituting the couple is zero.
(ii) Algebraic sum of the moments of the forces constituting the couple about any point is the same.
(iii) A single force cannot replace two equal and opposite forces whose lines of action are different.

### 4.5.3   Theorem on Couple

The algebraic sum of the moments of two forces forming a couple about any point in their plane is non-zero constant and equal to the moment of the couple.

Let each of the two forces forming the couple be *P* and *O* be any point in their plane. From *O*, draw perpendicular *OAB* to the lines of action of the forces meeting in *A* and *B* (Figure 4.9).

The algebraic sum of the moments of the forces about *O*

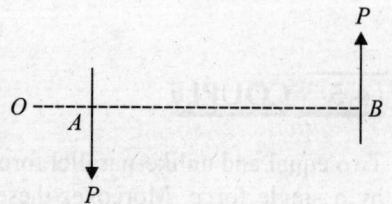

**Figure 4.9**   Proof of theorem in Section 4.5.3.

$$= P \times OB - P \times OA$$
$$= P (OB - OA)$$
$$= P \cdot AB$$

which is constant, i.e. independent of the position of *O* and it is equal to the moment of the couple.

**EXAMPLE 4.5** *ABCD* is a rectangle such that $AB = CD = a$ and $BC = DA = b$. Forces $P$ act along *AD* and *CB* and $Q$ act along *AB* and *CD* as shown in Figure 4.10. Prove that perpendicular distance $X$ between the resultant of the forces at $A$ and the resultant of forces at $C$ is:

$$X = \frac{Pa - Qb}{\sqrt{P^2 + Q^2}}$$

**Solution**   Considering Figure 4.10,
   Resultant of forces $P$ and $Q$ at $A$,

$$R_1 = \sqrt{P^2 + Q^2}$$

Similarly resultant of forces $P$ and $Q$ at $C$

$$R_2 = \sqrt{P^2 + Q^2}$$

$$\therefore \qquad R = R_1 = R_2$$

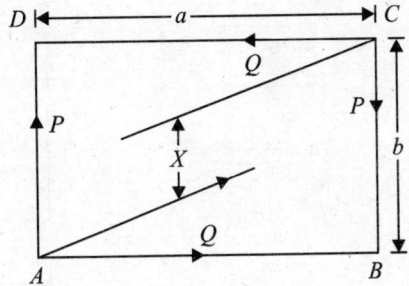

**Figure 4.10**   Visual of Example 4.5.

Taking the moment of $P$ about $C$,

$$M_1 = P \cdot a \ (+\text{ve})$$

Again moment of $Q$ about the same point

$$M_2 = -Qb \ (-\text{ve})$$

$$\therefore \qquad \text{Net moment } R = Pa - Qb \qquad\qquad\qquad \text{(i)}$$

Moment of the couple formed by the resultant $= R \times X$

$$R = \sqrt{P^2 + Q^2} \times X \qquad\qquad\qquad \text{(ii)}$$

Equating the moments of (i) and (ii)

$$Pa - Qb = \sqrt{P^2 + Q^2} \times X$$

$$\therefore \qquad X = \frac{Pa - Qb}{\sqrt{P^2 - Q^2}} \qquad \text{Proved.}$$

**EXAMPLE 4.6**   Four forces are completely represented by the sides *AB*, *BC*, *CD*, *DA* of a quadrilateral as shown in Figure 4.11. Show that they are equivalent to a couple consisting of two equal forces through $B$ and $D$.

**Solution**   Considering Figure 4.11,

   Forces $\overline{AB}$ and $\overline{BC}$ are equivalent to a force at $B$ represented in magnitude, direction and sense by $\overline{AC}$.

   Similarly forces $\overline{CD}$ and $\overline{DA}$ are equivalent to a force at $D$ represented in magnitude, direction and sense by $\overline{CA}$.

   Thus the four forces are equivalent to two equal, parallel and unlike forces at $B$ and $D$ and hence they are equivalent to a couple as shown in Figure 4.11.

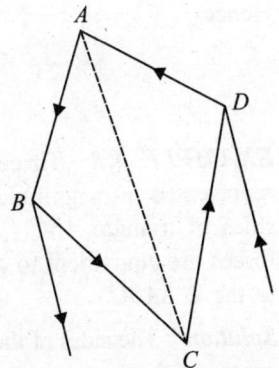

**Figure 4.11** Visual of Example 4.8.

***EXAMPLE 4.7***   $P$ and $Q$ are like-parallel forces. If a couple each of whose forces is $F$, and whose arm is $a$, in the plane of $P$ and $Q$, is combined with them, show that the resultant is displaced through a distance $\dfrac{Fa}{(P+Q)}$.

***Solution***   Figure 4.12 is drawn as per Example 4.7.

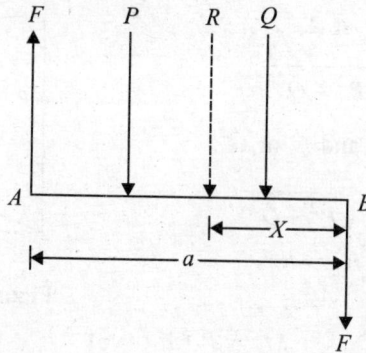

**Figure 4.12**   Visual of Example 4.7.

As shown in Figure 4.12, let the couple be constituted by force $F$ acting opposite with arm $a$.

Let $P$ and $Q$ be the like-parallel forces.

Let $R$ be the resultant of $P$ and $Q$ which is displaced at a distance $X$ from $B$ after the couple is introduced.

Then

$$R = P + Q + F - F$$
$$= (P + Q)$$

Taking moment of the forces about $B$,

$$R \times X = F \times a$$

or
$$(P + Q) \times X = Fa$$

Hence

$$X = \frac{Fa}{(P+Q)}$$

***EXAMPLE 4.8***   Three forces acting on a rigid body are represented in magnitude, direction and lines of action by three sides of triangle $ABC$ (Figure 4.13) taken in order. Prove that forces are equivalent to a couple whose moment $M$ is equivalent to the 2 $\triangle ABC$.

***Solution***   The sides of the triangles $AB$, $BC$ and $CA$ in Figure 4.14 represent the magnitudes and directions of the forces $P$, $Q$ and $R$ respectively.

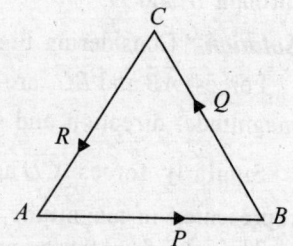

**Figure 4.13**   Visual of Example 4.8.

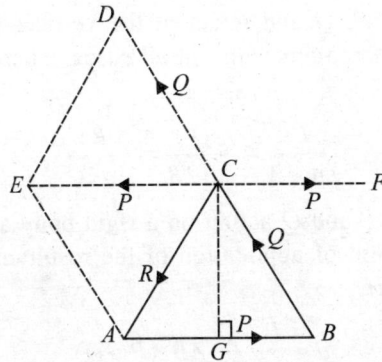

**Figure 4.14**  Solution of Example 4.8.

Now, line $EF$ is drawn through $C$ parallel to $AB$. Also extend $BC$ to $D$ such that $CD = CB$. Now apply two equal and opposite forces of $P$ magnitude at $C$, represented by $CE$ and $CF$.

Compute the parallelogram $CAED$ taking the two adjacent sides $CA$ and $CD$. Now from the law of parallelogram of forces, the diagonal $CE$, i.e. force $P$ represents the resultant of the two forces $Q$ and $R$.

The force represented by $CF$ equal in magnitude of $P$ will make the forces $Q$ and $R$ in equilibrium. Drop perpendicular $CG$ to $AB$ at $G$.

Now there are only two forces $AB$ equal to $P$ and $CE$ equal to $P$ which will form a couple and moment $M$ of this couple is:

$$M = P \times CG$$
$$= AB \times CG$$
$$= 2 \times \left(\frac{1}{2} AB \times CG\right) = 2 \times \left(\frac{1}{2} \text{base } AB \times \text{altitude } CG\right)$$

$\therefore$    $M = 2 \; \Delta ABC$     Proved.

## 4.6  CONCLUSION

Resultants of different parallel forces are presented and corresponding numerical examples to demonstrate the presentation are given. Couples are defined and movements of couples are shown with worked-out examples. A few problems are given in chapter-end exercises.

## EXERCISES

**4.1** Discuss the resultant of parallel forces in two different systems, i.e. like-parallel forces and unlike-parallel forces.

**4.2** Explain couple with its classification, characteristics and theorem with figures, where necessary.

**4.3** Three like-parallel forces $P$, $Q$ and $R$ act at the vertices of the triangle $ABC$. If their resultant passes through the circumcentre in all cases, whatever be the common direction of the forces, show that

$$\frac{P}{\sin 2A} = \frac{Q}{\sin 2B} = \frac{R}{\sin 2C}$$

**4.4** If two like-parallel forces $P$ and $Q$ acting on a rigid body at $A$ and $B$ be interchanged in position, show that the point of application of the resultant will be displaced along $AB$ through a distance $d$ where

$$d = \frac{P-Q}{P+Q} AB \ \ (P > Q)$$

**4.5** Find the like parallel forces acting at a distance of 24 cm apart whose resultant is 20 N and the line of action of one being at a distance of 60 cm from the given force.

(Answer   15 N and 5 N).

**4.6** A system of parallel forces is acting on a rigid bar $AB$ as shown in Figure 4.15. Reduce this system to a single force and a couple at $B$.

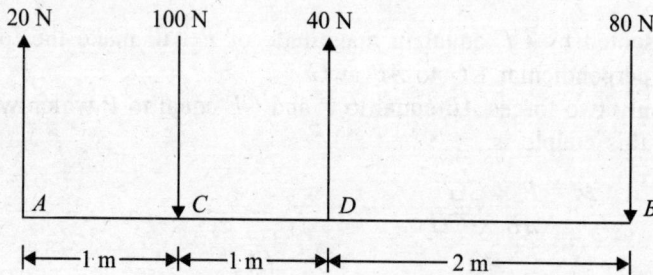

**Figure 4.15**   Visual of Exercise 4.6.

(Answer   $R = 120$ N, $M_B = 120$ Nm).

# CHAPTER 5

# Equilibrium of Forces

## 5.1  INTRODUCTION

A particle under the action of a number of forces is in equilibrium if the resultant of all the forces acting on the body is zero. Such a set of forces whose resultant is zero, is called *equilibrium of forces*. The force which causes the body to remain in equilibrium is called equilibrant. This equilibrant is equal and opposite to the resultant of the set of forces. In this chapter, the principles of equilibrium, force law of equilibrium, Lami's theorem of equilibrium, concept of free body diagram (FBD), types of support reactions and types of equilibrium are presented.

## 5.2  PRINCIPLE OF EQUILIBRIUM

The principle of equilibrium states that a stationary body under the action of a set of coplanar forces is in equilibrium if the following conditions are fulfilled.

$$\Sigma F = 0 \tag{5.1}$$
$$\Sigma M = 0 \tag{5.2}$$

Where $\Sigma F = 0$ indicates that the algebraic sum of all the forces is zero and $\Sigma M = 0$ indicates that the algebraic sum of moments of all the forces about any point in their plane is zero. Resolving the forces in horizontal and vertical directions, Eq. (5.1) may be written as

$$\Sigma F_x = 0 \tag{5.3}$$
$$\Sigma F_y = 0 \tag{5.4}$$

### 5.2.1  Two-Force Law or Principle

*It states that if a body is in equilibrium under the action of two forces, they must be equal, opposite and collinear.* This is shown in Figure 5.1.

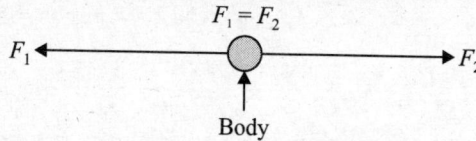

**Figure 5.1**   Two collinear, equal and opposite forces.

In Figure 5.1, force $F_1 = F_2$ and both are opposite and collinear and the body is in equilibrium. Equations (5.1) and (5.2) are fulfilled in the above system of two forces. But if two equal, opposite and parallel forces are acting on a body as shown in Figure 5.2, they do not fulfil all the conditions of equilibrium. Here $\Sigma F_x = 0$ and $\Sigma F_y = 0$ are satisfied as there is no horizontal forces and in vertical direction both the forces are equal and opposite. But the condition or equation $\Sigma M = 0$ is not satisfied as $\Sigma M = 0$ about the point $A$ is:

$$\Sigma M = F_2 \times BA + F_1 \times 0 = F_2 \times BA \neq 0$$

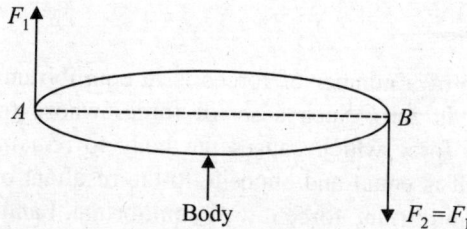

**Figure 5.2**   Two equal, opposite and parallel forces.

Thus the body will not be in equilibrium under the action of two equal, opposite and parallel forces but they produce a couple and moment of the couple which has already been discussed in Chapter 4.

## 5.2.2   Three-Force Law or Principle

*It states that if a body is in equilibrium under the action of three forces, then the resultant of the two forces must be equal, opposite and collinear with the third force.*

Consider three concurrent forces $F_1$, $F_2$ and $F_3$ in Figure 5.3. These three forces act at $A$ of the body. Let $R$ be the resultant of two forces $F_1$ and $F_2$. If $F_3$ is equal, opposite and collinear with the resultant $R$, then the body is in equilibrium. Thus three concurrent forces acting on a body at a point, set the body in equilibrium if the resultant of the two forces is equal, opposite and collinear to the resultant of the other two.

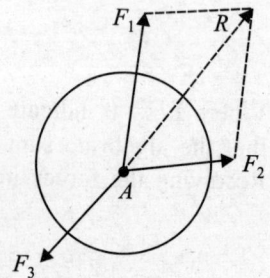

**Figure 5.3**   Three-force system.

## 5.2.3   Four-Force Law or Principle

*It states that if a body is in equilibrium under the action of four forces, then the resultant of any two forces must be equal, opposite and collinear to the resultant of the two forces.*

This is shown in Figure 5.4 in which resultant $R_1$ is equal, opposite and collinear with resultant $R_2$.

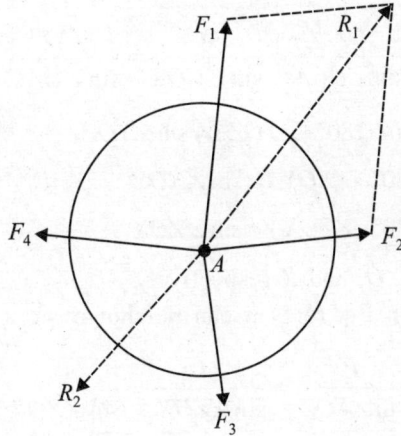

**Figure 5.4**   Four-force system.

## 5.3   LAMI'S THEOREM

The Lami's theorem is the application equilibrium of three forces acting on a body.

The Lami's theorem states that *if three forces acting at a point to be in equilibrium, then each is proportional to the sine of the angle between the other two.*

As shown in Figure 5.5, three forces $P$, $O$ and $R$ are acting at point $O$ along the lines $O\vec{X}, O\vec{Y}$ and $O\vec{Z}$  directions.

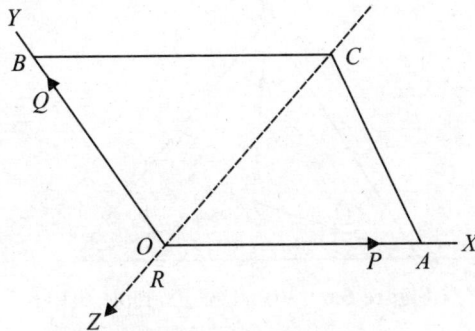

**Figure 5.5**   Lami's theorem.

It is required to prove that

$$\frac{P}{\sin \angle YOZ} = \frac{Q}{\sin \angle ZOX} = \frac{R}{\sin \angle XOY}$$

In Figure 5.5, cut $OA$ and $OB$ along $OX$ and $OY$ respectively equal in magnitude in a chosen scale. Complete the parallelogram $OACB$ and join $OC$. Now by the law of parallelogram of forces, the resultant of $P$ and $Q$ is represented by the diagonal $CO$. Since $P, Q, R$ are in equilibrium, the resultant of $P$ and $Q$ must be equal and opposite to $R$ so that $COZ^{\leftrightarrow}$ must be along the same line. Now from triangle $OAC$,

$$\frac{OA}{\sin \angle OCA} = \frac{AC}{\sin \angle COA} = \frac{CO}{\sin \angle OAC} \qquad (5.5)$$

but $\sin \angle OCA = \sin \angle COB = \sin (180° - \angle YOZ) = \sin \angle YOZ$

similarly $\sin \angle COA = \sin \angle (180° - \angle XOY) = \sin \angle XOY$

similarly $\sin \angle COA = \sin \angle (180° - \angle ZOX) = \sin \angle ZOX$

and $\overline{OA}, \overline{AC}, \overline{CO}$ represent $P$, $Q$, and $R$ respectively.

Hence putting the values in Eq. (5.5) it can be shown:

$$\frac{P}{\sin \angle YOZ} = \frac{Q}{\sin \angle ZOX} = \frac{R}{\sin \angle XOY} \qquad (5.6)$$

**EXAMPLE 5.1**   The three forces $P$, $Q$ and $R$ acting along $OA$, $OB$ and $OC$ (Figure 5.6) are equilibrium. If $O$ is the circumcentre of the triangle $ABC$, prove that

$$\frac{P}{a\left(\dfrac{b^2 + c^2 - a^2}{bc}\right)} = \frac{Q}{b\left(\dfrac{c^2 + a^2 - b^2}{ca}\right)} = \frac{R}{c\left(\dfrac{a^2 + b^2 - c^2}{ab}\right)}$$

where $a$, $b$, $c$ are the length of the sides of the triangle $BC$, $CA$, $AB$.

**Solution**   In Figure 5.6, $O$ is the circumcentre of triangle $ABC$.

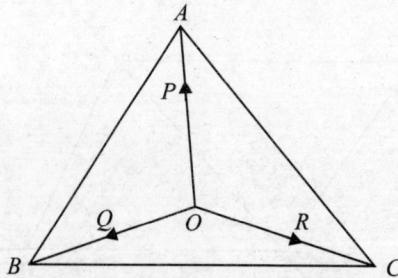

**Figure 5.6**   Visual of Example 5.1.

Hence $\angle BOC$ at centre $= 2\angle BAC$, similarly $\angle COA = 2\angle ABC$ and $\angle AOM = 2\angle ACB$
Since the forces $P$, $Q$, $R$ along $OA$, $OB$ and $OC$ are in equilibrium, Using Lami's theorem.

$$\frac{P}{\sin \angle BOC} = \frac{Q}{\sin \angle COA} = \frac{R}{\sin \angle AOB}$$

or

$$\frac{P}{\sin \angle 2A} = \frac{Q}{\sin \angle 2B} = \frac{R}{\sin \angle 2C}$$

or

$$\frac{P}{2\sin A \cos A} = \frac{Q}{2\sin B \cos B} = \frac{R}{2\sin C \cos C}$$

Now in the triangle $ABC$,

$$\frac{\sin A}{a} = \frac{\sin B}{B} = \frac{\sin c}{c} \quad \text{and} \quad \cos A = \frac{b^2 + c^2 - a^2}{2bc}, \text{etc.}$$

Hence from the above values after simplification, it can be shown as:

$$\frac{P}{a\left(\dfrac{b^2 + c^2 - a^2}{bc}\right)} = \frac{Q}{b\left(\dfrac{c^2 + a^2 - b^2}{ca}\right)} = \frac{R}{c\left(\dfrac{a^2 + b^2 - c^2}{ab}\right)}$$

**EXAMPLE 5.2** An electric light fixture weighing 20 N hangs from a point $C$ by two strings $AC$ and $BC$. $AC$ is inclined at 70° to the horizontal and $BC$ at 30° to the vertical as shown in Figure 5.7. Determine the forces in the strings $AC$ and $BC$.

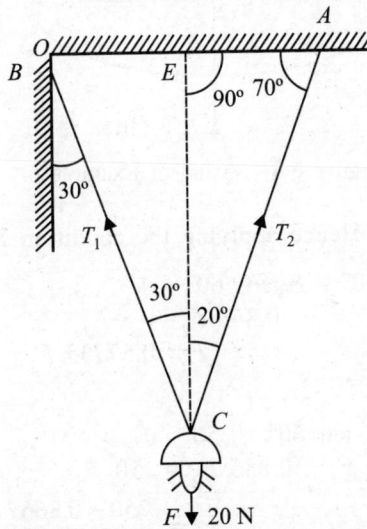

**Figure 5.7** Visual of Example 5.2.

**Solution** From the given angle of 70° and 30° (shown in Figure 5.7) in the Example, the angle $BCE$ and angle $ACE$ are worked out to be 30° and 20° respectively. The weight of light fixture is 20 N acting through $C$ vertically downwards. Let the tension of string $CB$ and $CA$ is $T_1$ and $T_2$. Using Lami's theorem at $C$,

$$\frac{20}{\sin \angle BCA} = \frac{T_1}{\sin \angle ACF} = \frac{T_2}{\sin \angle BCF}$$

Now

$$\angle BCA = 30° + 20° = 50°, \ \angle ACF = 180° - 20° = 160°, \ \angle BCF = 180° - 30° = 150°$$

$\therefore$ $$\frac{20}{\sin 50^\circ} = \frac{T_1}{\sin 160^\circ} = \frac{T_2}{\sin 150^\circ}$$

$\therefore$ $$T_1 = \frac{20 \times \sin 160^\circ}{\sin 50^\circ} = 8.9295 \text{ N} \qquad \text{(Answer)}$$

$\therefore$ $$T_2 = \frac{20 \times \sin 150^\circ}{\sin 50^\circ} = 13.054 \text{ N} \qquad \text{(Answer)}$$

**EXAMPLE 5.3** Three forces $F_1$, $F_2$ and $F_3$ are acting on a body as shown in Figure 5.8 and the body is in equilibrium. If the force $F_3$ is equal to 250 N, find the magnitude of $F_1$ and $F_2$.

**Solution** Consider Figure 5.8.

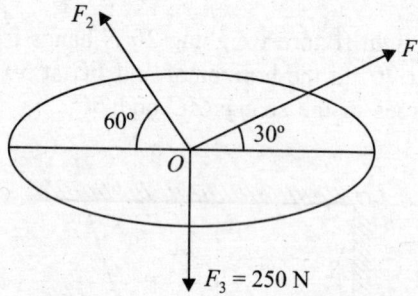

**Figure 5.8** Visual of Example 5.3.

The body is in equilibrium. Hence applying the condition $\Sigma F_x = 0$,

$$F_1 \cos 30^\circ - F_2 \cos 60^\circ = 0$$
$$0.866 \, F_1 = 0.5 \, F_2$$
or $$F_1 = 0.57735 \, F_2 \qquad \qquad \text{(i)}$$

(i) Applying the condition $\Sigma F_y = 0$,

$$F_1 \sin 30^\circ + F_2 \sin 60^\circ - 250 = 0$$
or $$0.5 \, F_1 + 0.886 \, F_2 = 250$$

or $$F_1 = \frac{250 - 0.866 \, F_2}{0.5} = 500 - 1.732 \, F_2 \qquad \text{(ii)}$$

Equating (i) and (ii),

$$0.57735 \, F_2 = 500 - 1.732 \, F_2$$
$\therefore$ $$F_2 = 216.511 \text{ N} \qquad \text{(Answer)}$$
and $$F_1 = 0.57735 \times 216.511 = 125 \text{ N} \qquad \text{(Answer)}$$

**Alternative method to solve Example 5.3**

The example can be solved by Lami's theorem as:

Using Lami's theorem to Figure 5.9,

$$\frac{F_1}{\sin 150^\circ} = \frac{F_2}{\sin 120^\circ} = \frac{250}{\sin 90^\circ}$$

or $$F_1 = 250 \frac{\sin 150°}{\sin 90°} = 250 \times \frac{0.5}{1} = 125 \text{ N} \qquad \text{(Answer)}$$

or $$F_2 = 250 \frac{\sin 120°}{\sin 90°} = 250 \times \frac{0.866025}{1} = 216.511 \text{ N} \qquad \text{(Answer)}$$

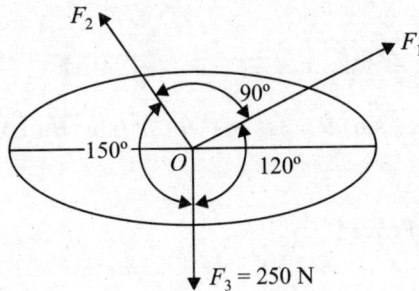

**Figure 5.9**   Visual of Example 5.3 to solve by Lami's theorem.

**EXAMPLE 5.4**  A heavy uniform rod of length a rests with one end against smooth vertical wall, the other end being tied to a point of the wall by a string of length *l* Prove that the rod will remain in equilibrium at an angle of $\theta$ to the wall, given by,

$$\cos^2 \theta = \frac{l^2 - a^2}{3a^2}$$

**Solution**   Figure 5.10 is drawn as per Example 5.4.

Let in Figure 5.10, AB be the length of rod a, BC be the string of length *l*.

The three forces that keep the rod in equilibrium are as follows:

Weight of the rod acting vertically downwards through its middle point G

The tension of the string acting along BC

The reaction of the smooth wall at A which must be normal to the wall

The wall is vertical and hence the reaction at A is horizontal.

These three forces keep the rod in equilibrium and hence they must meet at point O as shown in Figure 5.10.

Let $\theta$ be the inclination to the vertical at which the rod rests.

Now from Figure 5.10,

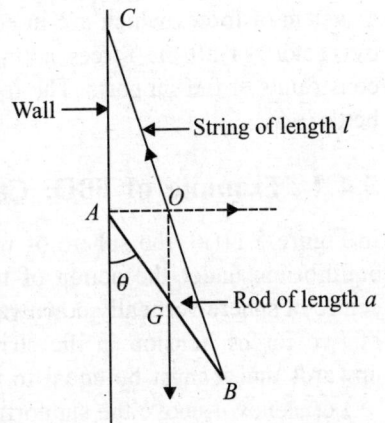

**Figure 5.10**   Visual of Example 5.4.

$$AO = AG \sin \theta = \frac{a}{2} \sin \theta, \quad OG = \frac{a}{2} \cos \theta$$

*GO* is parallel to *AC* through the midpoint *G* of *AB* and hence

$$CO = \frac{1}{2}BC = \frac{1}{2}l \quad \text{and} \quad AC = 2GO = a\cos\theta$$

Now from the triangle *ACO*,

$$CO^2 = AC^2 + AO^2$$

i.e.   $$\left(\frac{l}{2}\right)^2 = (a\cos\theta)^2 + \left(\frac{a}{2}\sin\theta^2\right)$$

or   $$l^2 = 4a^2\cos^2\theta + a^2\sin^2\theta = 3a^2\cos^2\theta + a^2(\sin^2\theta + \cos^2\theta) = 3a^2\cos^2\theta + a^2$$

or   $$l^2 - a^2 = 3a^2\cos^2\theta$$

∴   $$\cos^2\theta = \frac{l^2 - a^2}{3a^2} \quad \text{Proved}$$

## 5.4  FREE BODY DIAGRAM (FBD)

Statics deals with partially or completely constraint bodies which remain in rest under the action of forces. It is imagined that supports or constraints on the bodies are removed and they are replaced by reactions which they exert on the body. Thus it is seen that two kinds of forces act on a body to keep it at rest. They are the given or applied forces called the *active forces* and the reactions applied by the constraints or the supports called the *reactive forces*. In order to keep the body in equilibrium, it is necessary that both the active and reactive would form a system of forces which are in equilibrium. Free body diagram (FBD) is a diagram depicted by vectors of all the forces acting on the body, both active and reactive, after isolating the constraints or the supports. The following examples will clarify the concept of FBD in a much better way.

### 5.4.1  Example of FBD: Case I

In Figure 5.11(a), the sphere of weight *W* hangs from a support by a string. The sphere is in equilibrium under the action of two forces, i.e. weight *W* of the sphere acting through the centre of sphere vertically downwards and tension of the string and from the law of equilibrium of two forces, tension in the string *T* must act through the centre of the sphere vertically upwards and it must be equal to the weight *W*.

Let us now remove the supporting string and replace it by tension *T* that acts on the sphere as shown in Figure 5.11(b) in which the sphere is completely isolated from the support and the forces acting on it is shown by vectors. Therefore, Figure 5.11(b) is the free body diagram (FBD) for the sphere.

### 5.4.2  Example of FBD: Case II

In this case, a sphere of weight *W* is considered resting in contact with two smooth planes

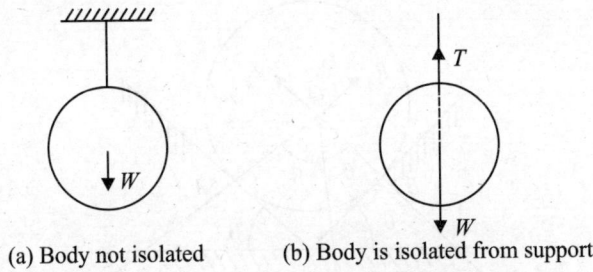

(a) Body not isolated        (b) Body is isolated from support

**Figure 5.11**   FBD when isolated from supports.

inclined at an angle of $\alpha$ and $\beta$ respectively to the horizon and intersecting in a horizontal line as shown in Figure 5.12(a). The sphere is at rest under the action of the following three forces:

Weight $W$ acting vertically downwards through the centre of the sphere.

The reaction of the smooth plane acting at $A$, perpendicular to this plane $Ox$ and passing through the centre of the sphere.

The reaction of the smooth plane acting at $B$, perpendicular to this plane $Oy$ and passing through the centre of the sphere.

Let us now remove the supporting planes $Ox$ and $Oy$ and replace them by supporting reactions $R_A$ and $R_B$ which exert on the sphere as shown by Figure 5.12(b). Thus Figure 5.12(b) shows that the sphere is isolated from the supports and forces acting on it are shown in vectors. Figure 5.12(b) is the free body diagram of the sphere.

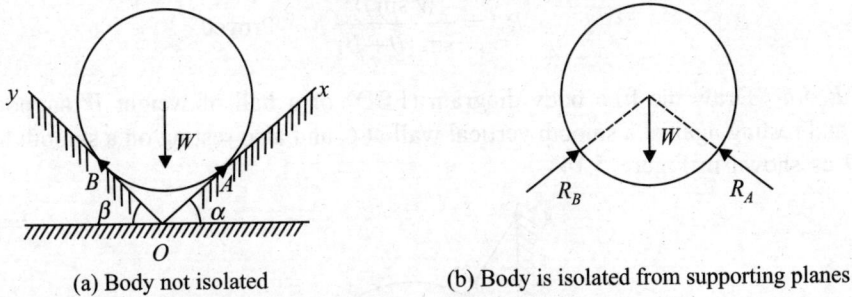

(a) Body not isolated        (b) Body is isolated from supporting planes

**Figure 5.12**   FBD when isolated from supports.

**EXAMPLE 5.5**   A spherical ball of weight $W$ rests in a triangular groove whose sides are inclined at angles $\theta$ and $\beta$ to the horizontal as shown in Figure 5.13. Show that

$$R_A = \frac{W \sin \theta}{\sin (\theta + \beta)} \quad \text{and} \quad R_B = \frac{W \sin \beta}{\sin (\theta + \beta)}$$

**Solution**   In Figure 5.13, The reactions $R_A$ and $R_B$ will be normal to the surface as shown. In order to keep the system in equilibrium, the three forces $W$, $R_A$ and $R_B$ mostly act through $O$ which is the centre of the sphere.

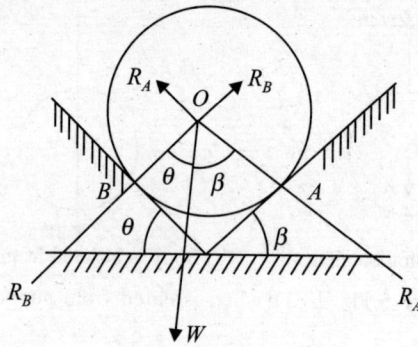

**Figure 5.13**  Visual of Example 5.5.

The sphere is in equilibrium under the three forces acting at $O$. Hence Lami's theorem can be applied as:

$$\frac{R_A}{\sin(180° - \theta)} = \frac{R_B}{\sin(180° - \beta)} = \frac{W}{\sin(\theta + \beta)}$$

or

$$\frac{R_A}{\sin\theta} = \frac{R_B}{\sin\beta} = \frac{W}{\sin(\theta + \beta)}$$

$\therefore$

$$R_A = \frac{W\sin\theta}{\sin(\theta + \beta)}$$

and

$$R_B = \frac{W\sin\beta}{\sin(\theta + \beta)} \qquad \text{Proved.}$$

**EXAMPLE 5.6**  Draw the free body diagram (FBD) of a ball of weight $W$ supported by a string $AB$ and resting against a smooth vertical wall at $C$ and also resting on a smooth horizontal floor at $D$ as shown in Figure 5.14

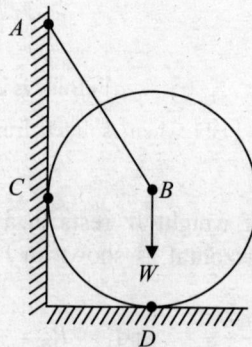

**Figure 5.14**  Visual of Example 5.6.

**Solution**  In order to draw the free body diagram of Figure 5.14, the ball should be isolated from vertical and horizontal supports and from the string $AB$.

Free body diagram of Figure 5.14 of the example is shown in Figure 5.15.

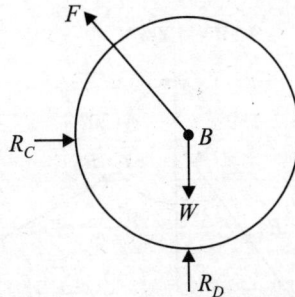

**Figure 5.15**   Free body diagram of Example 5.6.

In the free body diagram, reaction $R_C$ is normal to $AC$ of Figure 5.14 at $C$, $R_D$ is normal to the horizontal surface at $D$. $W$ is the weight acting vertically downwards through the centre of the ball. The force $F$ is in the direction of the string $BA$.

**EXAMPLE 5.7**   *A smooth circular cylinder of weight 1000 N and radius 10 cm rests on a right-angled groove whose sides are inclined at angle of 30° and 60° to the horizontal as shown in Figure 5.16. Determine the reactions $R_A$ and $R_C$ at the points of contact.*

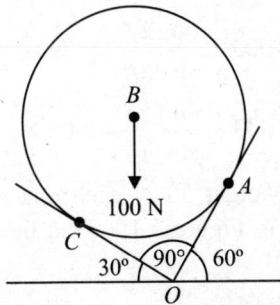

**Figure 5.16**   Visual of Example 5.7.

**Solution**   The free body diagram of the example is shown in Figure 5.17.

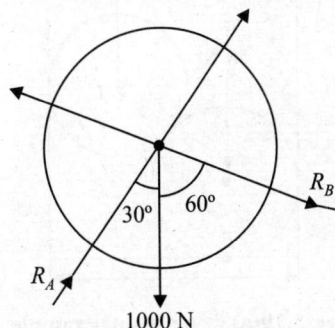

**Figure 5.17**   Free body diagram of Example 5.7.

The point of action of three forces $W$, $R_A$ and $R_B$ and the angles between the forces are calculated and all these are shown in Figure 5.18 in order to apply Lami's theorem.

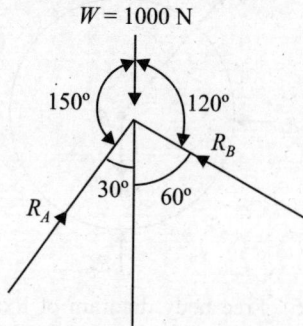

**Figure 5.18**   Angles between forces to apply Lami's theorem.

Now apply Lami's theorem,

$$\frac{W}{\sin 90°} = \frac{R_A}{\sin 120°} = \frac{R_B}{\sin 150°}$$

or

$$\frac{1000}{\sin 90°} = \frac{R_A}{\sin 120°} = \frac{R_B}{\sin 150°}$$

or

$$R_A = 1000 \times \frac{\sin 120°}{\sin 90°} = 886.025 \text{ N} \qquad \text{(Answer)}$$

or

$$R_B = 1000 \times \frac{\sin 150°}{\sin 90°} = 500 \text{ N} \qquad \text{(Answer)}$$

**EXAMPLE 5.8**   Two spheres of weight 50 N each and of radius 10 cm rest in horizontal channel of width 36 cm as shown in Figure 5.19. Find the reactions on the points of contact $A$, $B$ and $C$.

**Solution**   Considering Figure 5.19(a),

$$EF = 2 \times \text{radius} = 2 \times 10 = 20 \text{ cm and } FG = 16 \text{ cm}$$

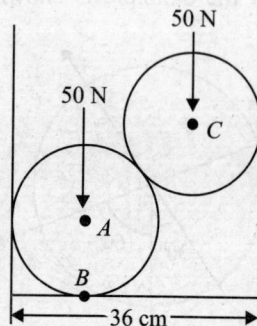

**Figure 5.19(a)**   Visual of Example 5.8.

Now from triangle *EFG*,

$$GE = \sqrt{EF^2 - FG^2} = \sqrt{20^2 - 16^2} = 12 \text{ cm}$$

and $$\cos\theta = \frac{GE}{EF} = \frac{12}{20} = \frac{3}{5} \quad \text{hence} \quad \sin\theta = \frac{4}{5}$$

Free body diagram of upper shpere is shown in Figure 5.19(b)

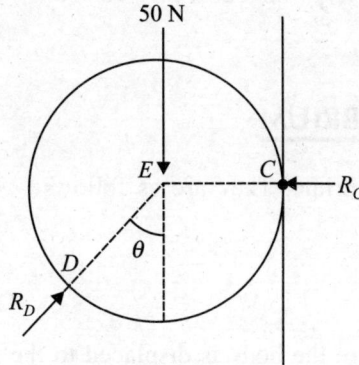

**Figure 5.19(b)**    Free body diagram of upper sphere.

for $$\Sigma F_x = 0, \quad R_D \sin\theta = RC \tag{i}$$
and for $$\Sigma F_y = 0, \quad R_D \cos\theta = 50 \text{ N}$$

$$R_D = \frac{50}{\cos\theta} = \frac{50 \times 5}{3} = \frac{250}{3} \text{ N}$$

Substituting the value of $R_D$ in (i),

$$\frac{250}{3} \times \frac{4}{5} = R_C \quad \therefore R_C = \frac{200}{3} = 66.667 \text{ N} \qquad \text{(Answer)}$$

Free body diagram of lower sphere is given in Figure 5.19(c)

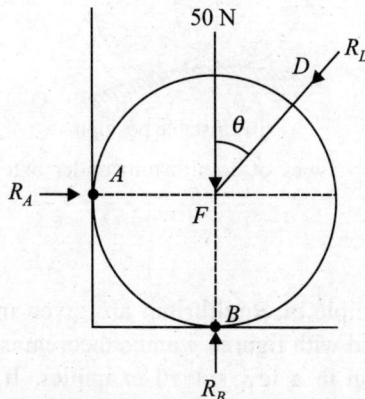

**Figure 5.19(c)**    Free body diagram of lower sphere.

For

$$\Sigma F_x = 0, \; R_A - R_D \sin\theta = 0$$

or

$$R_A = R_D \sin\theta = \frac{250}{3} \times \frac{4}{5} = \frac{200}{3} = 66.667 \text{ N} = R_C \qquad \text{(Answer)}$$

For

$$\Sigma F_y = 0, \; R_B = R_D \cos\theta + 50$$

or

$$R_B = \frac{250}{3} \times \frac{3}{5} + 50 = 50 + 50 = 100 \text{ N} \qquad \text{(Answer)}$$

## 5.5  TYPES OF EQUILIBRIUM

There are three types of equilibrium. They are as follows:

1. Stable equilibrium
2. Unstable equilibrium
3. Neutral equilibrium

Consider Figure 5.20(a). Here the body is displaced to the position shown by dotted lines by the application forces. If the body returns to its original position as shown, the body is said to have stable equilibrium.

Similarly in Figure 5.20(b), the body is seen in unstable position after displacement in which body is said to attain unstable equilibrium.

In Figure 5.20(c), the body remains at rest in its new position after the application of forces. Here the body is said to have neutral equilibrium.

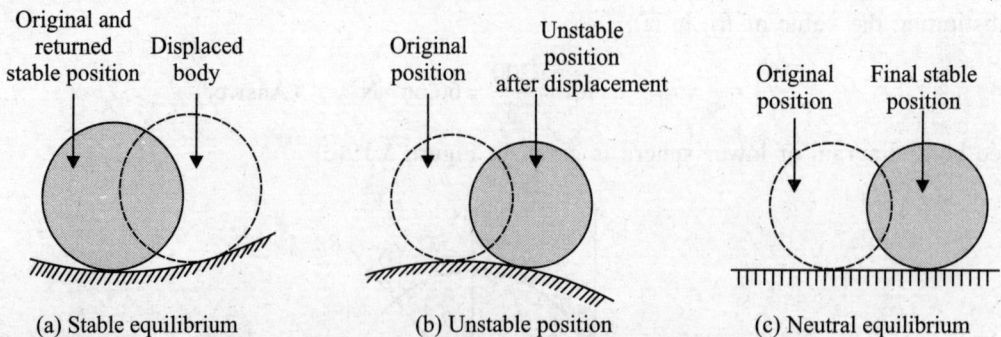

(a) Stable equilibrium          (b) Unstable position          (c) Neutral equilibrium

**Figure 5.20**   Types of equilibrium under external forces.

## 5.6  CONCLUSION

Equations that govern the principle of equilibrium are given in the beginning of the chapter. Then forces of laws are explained with figures. Lami's theorem is stated and proved. Application of this theorem has been shown in a few solved examples. It is a powerful tool of solving three-force law. Free body diagram (FBD) which is another tool for solving practical problems of equilibrium of body under the action of forces is presented and explained in two different situations. Examples and numerical problems are presented to understand the FBD. Finally, types of equilibrium of bodies under the action of forces are explained with sketches.

## EXERCISES

**5.1** State the principle of equilibrium and give analytical conditions of equilibrium.

**5.2** State Lami's theorem and explain it with figures.

**5.3** Explain and define free body diagram (FBD). Draw a free body diagram of a ball of weight $W$ placed on a horizontal surface.

**5.4** Two smooth spheres $A$ and $B$ each of diameter 40 cm and weight 20 N, rest in a horizontal channel having vertical walls and base width of 72 cm as shown in Figure 5.21.

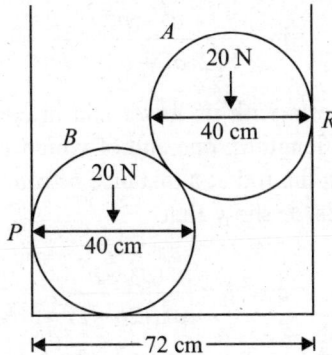

**Figure 5.21**   Visual of Exercise 5.4.

Find the reactions at $P$, $Q$ and $R$.

(Answer   26.667 N, 40 N, 26.667 N)

**5.5** Three cylinders weighing 100 N each and 80 mm diameter are placed in a horizontal channel as shown in Figure 5.22.

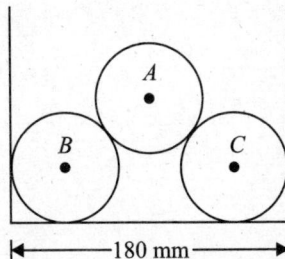

**Figure 5.22**   Visual of Exercise 5.5.

Determine the pressure exerted by
   (i) the cylinder $A$ on $B$ at the point of contact
  (ii) the cylinder $B$ on the base
 (iii) the cylinder $B$ on the wall at the point of contact

(Answer   64 N, 150 N, 40 N)

**5.6** A heavy uniform rod of length $2a$ rests in equilibrium, having one end against a smooth vertical wall and being placed upon a peg at a distance $b$ from the wall. Show that the inclination of the rod to the horizon is

$$\cos^{-1}\left(\frac{b}{a}\right)^{\frac{1}{3}}$$

**5.7** A uniform beam of length $l$ and weight $W$ hangs from a fixed point by two strings of length $a$ and $b$. Prove that the inclination of the rod to the horizon is

$$\sin^{-1}\frac{a^2-b^2}{l\sqrt{2(a^2+b^2)-l^2}}$$

**5.8** A uniform rod of length $2l$ rests with its lower end in contact with smooth vertical wall. It is supported by a string of length $a$, one end of which is fastened to a point in the wall and the other end to a point in the rod at a distance $b$ from the lower end. If the inclination of the string to the vertical is $\theta$, show that

$$\cos^2\theta=\frac{b^2(a^2-b^2)}{a^2l(2b-l)}$$

**5.9** A smooth sphere of weight $W$ is suspended in contact with a smooth vertical wall by a string fastened to a point on the surface, the other end being attached to a point in the wall. If the length is equal to the radius of the sphere, find the tension in the string and the reaction of the wall.

$$\left(\text{Answer}\quad\frac{2W}{\sqrt{3}},\frac{W}{\sqrt{3}}\right)$$

# CHAPTER 6

# Forces in Space: Introduction to Vector Algebra

## 6.1 INTRODUCTION

Vector quantities have both magnitude and direction. The terms such as force, velocity, acceleration, displacement, momentum, frequently used in engineering mechanics, are vector quantities. The laws of mechanics relate the above vector quantities. For examples, a force in space may have directions in three dimensions. Thus the magnitude and direction of this force can only correctly be represented by a vector. In order to handle these vector quantities in the laws of mechanics, knowledge of vector algebra is essential. Therefore, an attempt is made here to highlight at least the preliminary knowledge of vector algebra.

## 6.2 NOTATION, MAGNITUDE, ADDITION AND SUBSTRACTION OF VECTOR

A force which is vector quantity is graphically represented in vector algebra as bold letter **F** or $\vec{F}$ (Figure 6.1). In Figure 6.2, two vectors **A** and **B** are equal if they have the same magnitude and direction regardless of the position of their initial position and thus **A** = **B**. A vector having direction opposite to that of vector **A** but having the same magnitude, is denoted by –**A** (Figure 6.3). Magnitude of vector **F** is written as |**F**| = $F$, where $F$ is a positive quantity. The product of a vector **F** by a scalar $m$ is a vector $m$**F** with magnitude |$m$| times the magnitude of **A** and with direction the same or opposite to that of **A**, accordingly $m$ is positive or negative. If $m = 0$, $m$**A** is a *null vector.*

To find sum of resultant of two vectors, consider two vectors **A** and **B** shown in Figure 6.1. To add these two vectors, draw a line $RQ$ parallel to vector **B** and equal in magnitude in suitable scale. Through $R$ draw a line parallel to vector **A** and equal in magnitude and in the same scale. Join $PQ$ which is the sum of vectors **A** and **B** which may be written as another vector **C** = **A** + **B** as shown in Figure 6.4. In case of subtraction, consider two vectors **A** and **B** in Figure 6.4. Now the line $PR$ is equal in magnitude in the direction of vector **A**. Vector **B** is drawn through point $R$ equal in magnitude but opposite in direction.

Resultant vector $C = A - B$ as shown in Figure 6.5.

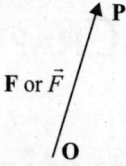

**Figure 6.1** Graphical representation of vector **OP**.

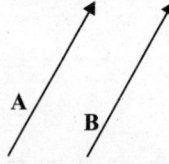

**Figure 6.2** Two equal vectors **A** and **B**.

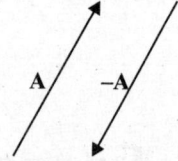

**Figure 6.3** Two equal vectors of opposite direction.

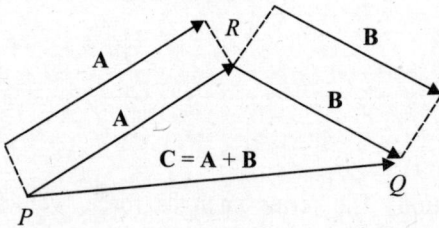

**Figure 6.4** Addition of vectors.

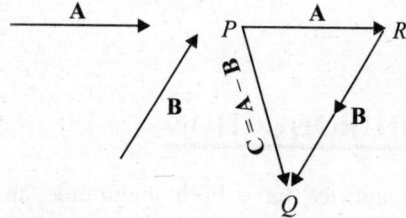

**Figure 6.5** Subtraction of vectors.

## 6.3 LAWS OF VECTOR ALGEBRA

If **A, B** and **C** are vectors and $m$ and $n$ are scalars, then
Cumulative law of addition is

$$A + B = B + A$$

Associative law of addition

$$(A + B) + C = A + (B + C)$$

Cumulative law of multiplication

$$mA = Am$$

Associative law of multiplication

$$m(nA) = (mn)A$$

Distributive laws

$$(m + n)A = mA + nA$$
$$m(A + B) = mA + mB$$

The above laws enable to treat vector equations in the same way as ordinary algebraic equations.
For example, if $A + B = C$ then by transposing $A = C - B$.

## 6.4 RECTANGULAR UNIT VECTORS i, j, k

Rectangular unit vectors are which have the directions of positive $x$, $y$ and $z$ axes of a three-dimensional rectangular coordinate system. They are denoted by **i, j** and **k** respectively as shown in Figure 6.6.

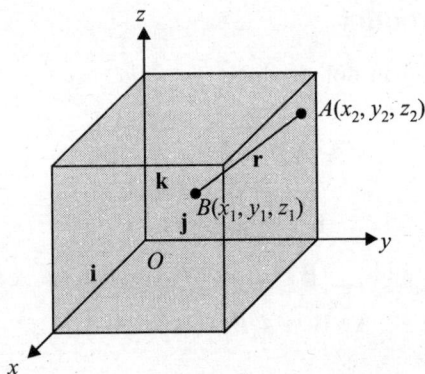

**Figure 6.6**   Rectangular unit vectors **i**, **j**, **k**.

If $F_1$, $F_2$ and $F_3$ are the rectangular components of force vector **F** with respect to origin $O$ in $x$, $y$ and $z$ directions respectively, then sum of the resultant of **F** can be written as:

$$\mathbf{F} = \mathbf{i}F_1 + \mathbf{j}F_2 + \mathbf{k}F_3 \text{ and magnitude is } F = |F| = \sqrt{F_1^2 + F_2^2 + F_3^2}$$

The position vector of a point $A$ with coordinates $(x_2, y_2, z_2)$ with respect to a point $B$ $(x_1, y_1, z_1)$ is given by:

$$\mathbf{r} = (x_2 - x_1)\,\mathbf{i} + (y_2 - y_1)\,\mathbf{j} + (z_2 - z_1)\,\mathbf{k}$$

## 6.5   MULTIPLICATION OF VECTORS

Vector multiplication is a bit more complicated. There are two kinds of multiplication which are useful in mechanics. They are a *dot* or *scalar product* and a *cross* or *vector product*.

### 6.5.1   The Dot or Scalar Product

The dot or scalar product of vectors **A** and **B** is denoted by:

$$\mathbf{A} \cdot \mathbf{B} = AB \cos\theta = |A|\,|B|\cos\theta \text{ when } 0 \le \theta \le \pi$$

Here $|A|$ and $|B|$ denote the magnitudes of vector **A** and **B** and $\theta$ is the angle between the two vectors. The dot product $AB \cos\theta$ or is a scalar quantity. If $\mathbf{A} \cdot \mathbf{B} = 0$, then **A** is perpendicular to **B** or the vectors are orthogonal since 90° is zero.

The practical example of dot product is the work done $W$ by a force factor **F** and displacement vector **D** as shown in Figure 6.7 and mathematically it is shown as:

$$W = \mathbf{F} \cdot \mathbf{D} = FD \cos\theta = |F|\,|D|\cos\theta \text{ when } 0 \le \theta \le \pi$$

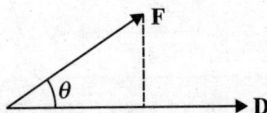

**Figure 6.7**   Work done as a product of vector force and displacement.

### 6.5.2   Laws of Dot Product

The following laws are valid in dot product.

$$\mathbf{A} \cdot \mathbf{B} = \mathbf{B} \cdot \mathbf{A}$$
$$\mathbf{A} \cdot (\mathbf{B} + \mathbf{C}) = \mathbf{A} \cdot \mathbf{B} + \mathbf{A} \cdot \mathbf{C}$$
$$\mathbf{i} \cdot \mathbf{i} = \mathbf{j} \cdot \mathbf{j} = \mathbf{k} \cdot \mathbf{k} = 1$$
$$\mathbf{i} \cdot \mathbf{j} = \mathbf{j} \cdot \mathbf{k} = \mathbf{k} \cdot \mathbf{i} = 0$$

If $\mathbf{A} = A_1\mathbf{i} + A_2\mathbf{j} + A_3\mathbf{k}$    and    $\mathbf{B} = B_1\mathbf{i} + B_2\mathbf{j} + B_3\mathbf{k}$

then
$$\mathbf{A} \cdot \mathbf{B} = A_1B_1 + A_2B_2 + A_3B_3 \tag{6.1}$$

### 6.5.3   Cross Product of Vectors

The cross or vector product of **A** and **B** is vector **C** and is denoted by:

$$\mathbf{A} \times \mathbf{B} = \mathbf{C} = \begin{vmatrix} \mathbf{i} & \mathbf{j} & \mathbf{k} \\ A_x & A_y & A_z \\ B_x & B_y & B_z \end{vmatrix}$$

Here **i**, **j**, **k** are unit vectors in $x$, $y$ and $z$ axes of three-dimensional rectangular coordinate system, $A_x$, $A_y$ and $A_z$ are components of vector **A**, and $B_x$, $B_y$ and $B_z$ are components of vector **B** in $x$, $y$ and $z$ directions respectively. The magnitude of $\mathbf{A} \times \mathbf{B} = \mathbf{C}$ is the area of the parallelogram with sides **A** and **B**. The direction of **C** is normal to the it plane of **A** and **B** and points in the direction of movement of a right-handed screw turning from **A** towards **B** as shown in Figure 6.8. Again if the vector **A** is a force **F** and vector **B** is a distance vector **r** with respect to $O$, centre of $x$, $y$ and $z$ coordinate system, then $\mathbf{F} \times \mathbf{r} = \mathbf{M}$, **M** is the moment in vector of the force vector **F** about $O$ which is at distance vector **r**. Thus moment **M** of force **F** about distance **r** is:

$$\mathbf{M} = \mathbf{F} \times \mathbf{r} = \begin{vmatrix} \mathbf{i} & \mathbf{j} & \mathbf{k} \\ x & y & z \\ F_x & F_y & F_z \end{vmatrix}$$

$$\mathbf{M} = \mathbf{i}\,(yF - zF) + \mathbf{j}\,(zF - xF) + \mathbf{k}\,(xF - yF)$$

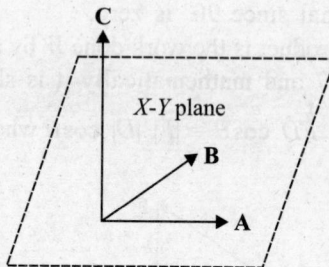

**Figure 6.8**   Orientation of vector cross product $\mathbf{A} \times \mathbf{B} = \mathbf{C}$.

Here $F_x$, $F_y$ and $F_z$ are components of the force **F** along $x$, $y$ and $z$ coordinates, i.e.

$$\mathbf{F} = \mathbf{i}F_x + \mathbf{j}F_y + \mathbf{k}F_z$$

Similarly

$$\mathbf{r} = \mathbf{i}x + \mathbf{j}y + \mathbf{k}z$$

### 6.5.4  Laws of Cross Product

The following laws of cross product are valid.

$$\mathbf{A} \times \mathbf{B} = -\mathbf{B} \times \mathbf{A}$$
$$\mathbf{A} \times (\mathbf{B} + \mathbf{C}) = \mathbf{A} \times \mathbf{B} + \mathbf{A} \times \mathbf{C}$$
$$\mathbf{i} \times \mathbf{i} = \mathbf{j} \times \mathbf{j} = \mathbf{k} \times \mathbf{k} = 0$$
$$\mathbf{i} \times \mathbf{j} = \mathbf{k}, \ \mathbf{j} \times \mathbf{k} = \mathbf{i}, \ \mathbf{k} \times \mathbf{i} = \mathbf{j}$$

## 6.6  GRADIENT, DIVERGENCE AND CURL

The vector differential operator written as $\nabla$ is defined by:

$$\nabla = \frac{\delta}{\delta x}\mathbf{i} + \frac{\delta}{\delta y}\mathbf{j} + \frac{\delta}{\delta z}.\mathbf{k} = \mathbf{i}\frac{\delta}{\delta x} + \mathbf{j}\frac{\delta}{\delta y} + \mathbf{k}\frac{\delta}{\delta z}$$

This vector operator has the properties analogous to those of ordinary vectors. It is useful to define three quantities which are known as the *gradient*, the *divergence* and the *curl*.

### 6.6.1  Gradient ($\nabla$)

If $\phi(x, y, z)$ defines a differential scalar field, then gradient of $\phi$ written as $\nabla\phi$ or grad of $\phi$, is defined by:

$$\nabla\phi = \left(\frac{\delta}{\delta x}\mathbf{i} + \frac{\delta}{\delta y}\mathbf{j} + \frac{\delta}{\delta z}\mathbf{k}\right)\phi$$

$$= \mathbf{i}\frac{\delta\phi}{\delta x} + \mathbf{j}\frac{\delta\phi}{\delta y} + \mathbf{k}\frac{\delta\phi}{\delta z}$$

It should be noted $\nabla\phi$ defines a vector field.

### 6.6.2  The Divergence

If $\mathbf{F}(x, y, z) = F_x\mathbf{i} + F_y\mathbf{j} + F_z\mathbf{k}$ be differentiable at each point $(x, y, z)$ in a certain region of space, i.e. **F** defines differential force vector field, then divergence of **F** is written as $\nabla \cdot \mathbf{F}$ or div **F** and it is defined as:

$$\nabla \cdot \mathbf{F} = \left(\frac{\delta}{\delta x}\mathbf{i} + \frac{\delta}{\delta y}\mathbf{j} + \frac{\delta}{\delta z}\mathbf{k}\right) \cdot \left(F_x\mathbf{i} + F_y\mathbf{j} + F_z\mathbf{k}\right)$$

$$= \frac{\delta F_x}{\delta x} + \frac{\delta F_y}{\delta y} + \frac{\delta F_z}{\delta z}$$

### 6.6.3 The Curl

Let $V(x, y, z)$ be a differential vector field. Then the curl or rotation of $V$ is written as $\nabla \times V$, curl $V$ or rot $V$ which is defined by:

$$\nabla \times V = \left( \frac{\delta}{\delta x} i + \frac{\delta}{\delta y} j + \frac{\delta}{\delta z} k \right) \times \left( V_x i + V_y j + V_z k \right)$$

$$= \begin{vmatrix} i & j & k \\ \dfrac{\delta}{\delta x} & \dfrac{\delta}{\delta y} & \dfrac{\delta}{\delta z} \\ V_x & V_y & V_z \end{vmatrix}$$

$$= i \begin{vmatrix} \dfrac{\delta}{\delta y} & \dfrac{\delta}{\delta z} \\ V_y & V_z \end{vmatrix} - j \begin{vmatrix} \dfrac{\delta}{\delta x} & \dfrac{\delta}{\delta z} \\ V_x & V_z \end{vmatrix} + k \begin{vmatrix} \dfrac{\delta}{\delta x} & \dfrac{\delta}{\delta y} \\ V_x & V_y \end{vmatrix}$$

$$= i \left( \frac{\delta V_z}{\delta_y} - \frac{\delta V_y}{\delta_z} \right) + j \left( \frac{\delta V_x}{\delta_z} - \frac{\delta V_z}{\delta_x} \right) + k \left( \frac{\delta V_y}{\delta_x} - \frac{\delta V_x}{\delta_y} \right)$$

***EXAMPLE 6.1***   An automobile travels 3 km to north, then 4 km towords north-east. Represent these displacements graphically and determine the resultant displacement and its direction.

***Solution***   In Figure 6.9, vector **OP** or **A** represents displacement of 3 km to north. Vector **PQ** or **B** represents displacement of 4 km north-east. The vector **OQ** or **C** = **A** + **B** represents the resultant displacement or sum of vectors **A** and **B**. This is called *triangle law of vector addition*. The resultant vector **OQ** can also be obtained by constructing the diagonal of the parallelogram *OPQR* with vectors **OP** = **A** and **OR** equal to vector **PQ** or **B** as sides of the parallelogram. This is called the *parallelogram law of vector addition*.

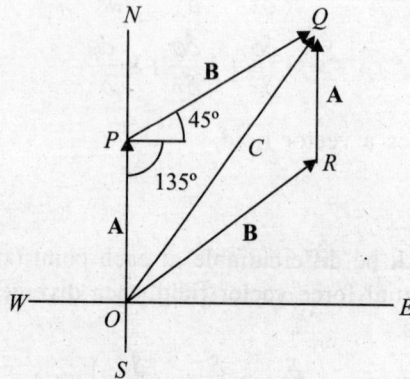

**Figure 6.9**   Visual of Example 6.1.

In the triangle *OPQ*, denote magnitude of **A**, **B**, and **C** by *A*, *B* and *C* respectively.

Using the law of cosine from trigonometry:

$$C^2 = A^2 + B^2 - 2AB \cos \angle OPQ$$
$$C^2 = 3^2 + 4^2 - 2 \times 3 \times 4 \times \cos 135°$$
$$C^2 = 9 + 16 - 24 \times (-0.7071)$$
$$C^2 = 41.9704$$
$$\therefore \qquad C = 6.4784 \text{ km}$$

Thus $C = A + B = 6.4784$ km      (Answer)

By the law of sine,

$$\frac{A}{\sin \angle OQP} = \frac{C}{\sin \angle OPQ}$$

or $\qquad \sin \angle OQP = \dfrac{A \sin \angle OPQ}{C} = \dfrac{3 \times \sin 135°}{6.4784} = \dfrac{3 \times (0.7071)}{6.4784} = 0.32744$

$\therefore \qquad \angle OQP = 19°6'48'' = \angle QOR$

Hence the direction of vector **OQ** $= (45° + 19°6'48'') = 64.°6'48''$ north-east      (Answer)

**EXAMPLE 6.2**   A force vector **A** kN causes a displacement of **B** m. If $A = 5i - 4j - 3k$ and $B = 4i + 3j - 2k$, determine the work done.

**Solution**   Force vector

$$A = 5i - 4j - 3k$$

where

$$A_x = 5, \ A_y = -4, \ A_z = -3$$

Displacement vector

$$B = 4i + 3j - 2k$$

where

$$B_x = 4, \ B_y = 3, \ B_z = -2$$

Dot product of these two vectors **A** and **B** gives the work done $W$ which is a scalar quantity. Using Eq. (6.1),

$$A \cdot B = \text{work done } W = (5i - 4j - 3k) \cdot (4i + 3j - 2k)$$
$$W = A_x B_x + A_y B_y + A_z B_z$$
$$W = 5 \times 4 + [(-4) \times 3] + [(-3) \times (-2)]$$
$$W = 20 - 12 + 6$$
$$W = 14 \text{ kN-m}\qquad \text{(Answer)}$$

**EXAMPLE 6.3**   A force vector **A** is equal to $(12i + 7j - 10k)$ N. The point of application of the force moves from the point $(3i + 2k)$ m to the point $(5i - 2j - 5k)$ m Find the work done by the force.

**Solution**   Force vector

$$A = 12i + 7j - 10k$$
$$\text{Initial point} = 3i + 2k$$
$$\text{Final point} = 5i - 2j - 5k$$

Let the distance vector

$$\mathbf{B} = \text{Final point} - \text{initial point}$$
$$\mathbf{B} = (5\mathbf{i} - 2\mathbf{j} - 5\mathbf{k}) - (3\mathbf{i} + 2\mathbf{k})$$
$$\mathbf{B} = 2\mathbf{i} - 2\mathbf{j} - 7\mathbf{k}$$

Work done $W$ is the dot product of force $\mathbf{A}$ and the displacement or distance vector $\mathbf{B}$. Hence

$$W = \mathbf{A} \cdot \mathbf{B} = (12\mathbf{i} + 7\mathbf{j} - 10\mathbf{k}) \cdot (2\mathbf{i} - 2\mathbf{j} - 7\mathbf{k})$$
$$W = 12 \times 2 + 7 \times (-2) + (-10) \times (-7)$$
$$W = 24 - 14 + 70 = 80 \text{ kN} \qquad \text{(Answer)}$$

**EXAMPLE 6.4**   Prove that the area of a parallelogram with sides $\mathbf{A}$ and $\mathbf{B}$ is $|\mathbf{A} \times \mathbf{B}|$.

**Solution**   Consider Figure 6.10,

$$\text{Area of a parallelogram} = \frac{1}{2}|\mathbf{B} + \mathbf{B}| \times h$$
$$= |\mathbf{B}| \times h$$
$$= |\mathbf{B}| \, |\mathbf{A}| \, \sin\theta$$
$$= |\mathbf{A} \times \mathbf{B}|$$

It should be noted that area of trinagle with sides $\mathbf{A}$ and $\mathbf{B}$

$$= \frac{1}{2}|\mathbf{A} \times \mathbf{B}|$$

Hence area of the parallelogram $= |\mathbf{A} \times \mathbf{B}|$   Proved.

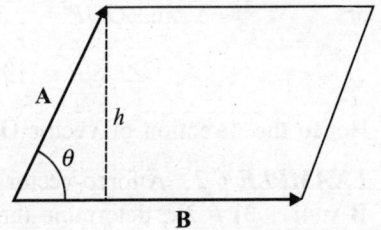

**Figure 6.10**   Visual of Example 6.4.

## 6.7   CONCLUSION

Vector algebra, which had its beginning in the middle of the nineteenth century, has become an essential part in recent years of the mathematical background required for engineers, mathematicians, physicists, and scientists. The presentation in this chapter has shown that analysis of forces, displacements, momentums, velocity, accelerations in space can be best understood with the help of vector algebra. It has now wide applications in mechanics of solids, fluid mechanics, i.e. all kinds of engineering mechanics. Since the dot product, cross product of vectors, etc. has their practical application in mechanics, an introduction on this is presented in this chapter for this first course of engineering mechanics.

## EXERCISES

**6.1** An aircraft travels 200 km to west and then 150 km 60° north of west. Determine the resultant displacement analytically.

(Answer   Magnitude = 304.138 km, direction = 25°17′)

**6.2** Determine the resultant of the following displacements: **A** is 20 km 30° south of east; **B** is 50 km to west; **C** is 40 km north-east; **D** is 30 km 60° south of west.

(Answer   Magnitude = 20.9 km, direction = 21°39′ south of west)

**6.3** Three coplanar forces act on an object $Q$ as shown in Figure 6.11. Determine the force needed to prevent $Q$ from moving.

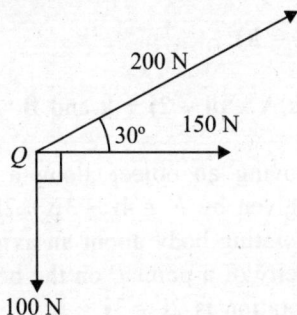

**Figure 6.11**  Visual of Exercise 6.3.

(Answer   323 N directly opposite to 150 N)

**6.4** If *ABCDEF* are the vertices of a regular hexagon, find resultant of the forces represented by the vectors **AB, AC, AD, AE** and **AF**. (Answer   3 **AD**)

**6.5** A weight of 100 N is suspended from the centre of a rope as shown in Figure 6.12. Determine the tension $T$ in the rope.

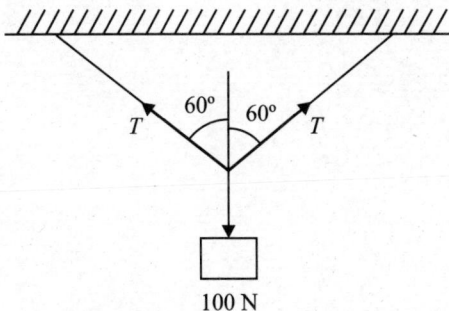

**Figure 6.12**  Visual of Exercise 6.5. (Answer   100 N)

**6.6** *ABCD* is parallelogram with $P$ and $Q$ be the midpoints of sides $BC$ and $CD$ respectively as shown in Figure 6.13. Prove that $AP$ and $AQ$ trisect the diagonal $BD$ at points $E$ and $F$.

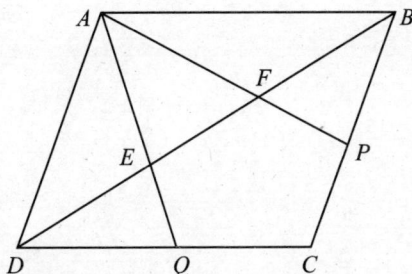

**Figure 6.13**  Visual of Exercise 6.6.

**6.7** Evaluate

(a) $\mathbf{k} \cdot (\mathbf{i} + \mathbf{j})$

(b) $(\mathbf{i} - 2\mathbf{k}) \cdot (\mathbf{i} + 2\mathbf{k})$

(c) $(2\mathbf{i} - \mathbf{j} + 2\mathbf{k}) \cdot (3\mathbf{i} + 2\mathbf{j} - \mathbf{k})$

(Answer   (a) 0;  (b) –6;  (c) 1)

**6.8** For what value of a vectors $\mathbf{A} = a\mathbf{i} - 2\mathbf{j} + \mathbf{k}$ and $\mathbf{B} = 2a\mathbf{i} + a\mathbf{j} - 4\mathbf{k}$ are perpendicular?

(Answer   $a = 2, -1$)

**6.9** Find the work done in moving an object along a straight line from (3, 2, –1) to (2, –1, 4) in a force field given by $F = 4\mathbf{i} - 3\mathbf{j} + 2\mathbf{k}$.   (Answer   15)

**6.10** The angular velocity of a rotating body about an axis of rotation is given by $\omega = 4\mathbf{i} + \mathbf{j} - 2\mathbf{k}$. Find the linear velocity of a point $P$ on the body whose position vector relative to a point on the axis of rotation is $2\mathbf{i} - 3\mathbf{j} + \mathbf{k}$.   (Answer   $-5\mathbf{i} - 3\mathbf{j} - 14\mathbf{k}$)

# CHAPTER 7

# Analysis of Forces in Perfect Frames

## 7.1 INTRODUCTION

Frame is a structure made of several members as shown in Figure 7.1. The members may be angles of irons or channel sections. They are riveted or welded on the joints. For analysis of forces in the frames, these joints are assumed to be pin-jointed or hinged. The determination of forces in the members of the frame when subjected to external loads is very important in the field of engineering. With the application of principles of statics, forces in different members can be worked out quite easily.

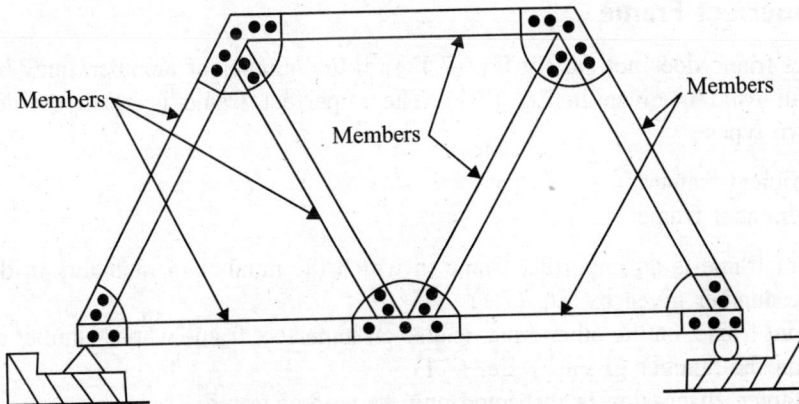

**Figure 7.1** A simple frame or truss.

## 7.2 CLASSIFICATION OF FRAMES

The frames are classified as *perfect* and *imperfect* frames.

## 7.2.1   Perfect Frame

A perfect frame is composed of the members which are just sufficient to keep it in equilibrium without any change of its shape. The simplest perfect frame is triangular in shape as shown in Figure 7.2(a) which has 3 joints and 3 members. Figure 7.2(b) is another frame with 4 joints and 5 members.

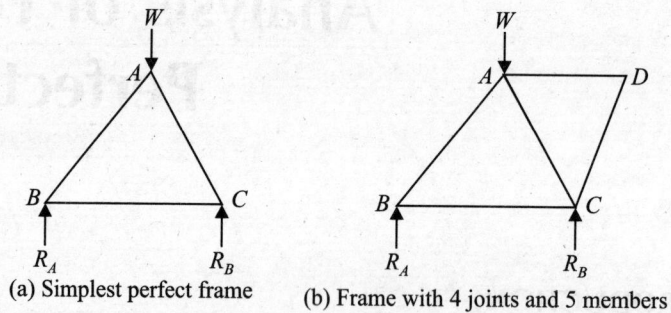

(a) Simplest perfect frame          (b) Frame with 4 joints and 5 members

**Figure 7.2**   Frame with different number of joints and members.

Again if Figure 7.1 is considered, it has 5 joints and 7 members. Thus it is seen that in every additional joint to a triangular frame, two members are required. It can be written that the number $n$ in perfect frame is related to joints $j$ as:

$$n = (2j - 3) \tag{7.1}$$

## 7.2.2   Imperfect Frame

An imperfect frame does not satisfy Eq. (7.1), i.e. the number of members may be less than or more than what is given in Eq. (7.1). The imperfect frame is again divided into the following two types:

(i) Deficient frame
(ii) Redundant frame

A deficient frame is an imperfect frame in which the number of members in the frame is less than the number given by Eq. (7.1)

A redundant frame, on the other hand, is also an imperfect frame whose number of members are more than the number given by Eq. (7.1).

In this chapter, discussion is restricted only to perfect frame.

## 7.3   ASSUMPTIONS MADE IN THE DETERMINATION OF FORCES IN MEMBERS

The following assumptions are made in the determination of the forces in the members.

(i) The frame is perfect
(ii) The frame is loaded only on the joints

(iii) All members are pin-jointed

(iv) Self-load of the frame is usually neglected unless it is mentioned.

## 7.4    METHODS OF FINDING THE FORCES IN MEMBERS

There are two methods of finding the forces in the members.

(i) Analytical method

(ii) Graphical method

Analytical method is again divided into two parts. They are:

(i) Method of sections or method of moments

(ii) Method of joints

An analysis in this chapter will be made by analytical method as graphical methods are not preferred with the advent of computer application.

## 7.5    METHODS OF SECTIONS

If the number of members from which the forces are to be found out is not many, this method is particularly suitable.

In Figure 7.3(a), a simple frame $ABC$ is considered with a load $W$ at $C$. To find forces in members of the frame, a section 1–1 is taken as shown in Figure 7.3(b). While drawing a section line, care is to be taken not to cut more than three members, in which the forces are unknown. A part of the frame on the left side of the frame is taken out as a free body diagram in Figure 7.3(c). This free body diagram is in equilibrium under the action of external forces. The unknown forces are obtained by applying the equation of equilibrium, i.e.

$$\Sigma M = 0 \qquad\qquad (7.2)$$

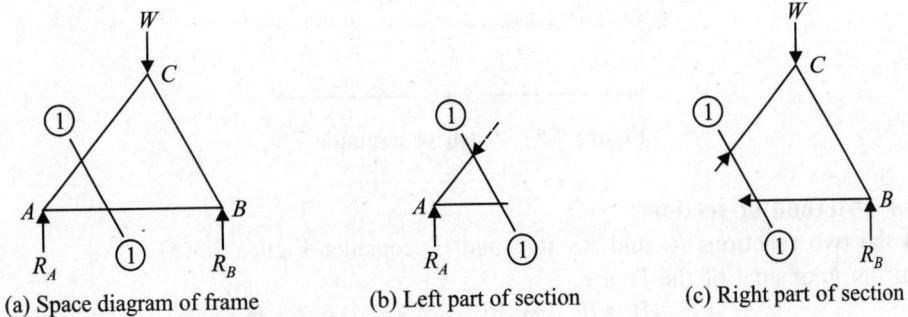

| (a) Space diagram of frame | (b) Left part of section | (c) Right part of section |

**Figure 7.3**   Determination of forces by method of section.

## 7.6    METHOD OF JOINTS

In the method of joints, every joint is treated separately as free body diagram which is in equilibrium. In selecting the joints for calculation, care is to be taken that at any instant, the joint should not contain more than two members in which the forces are unknown.

In Figure 7.4, joints $A$, $C$ and $B$ are considered separately in Figure 7.4(b), 7.4(c) and 7.4(d) respectively. The unknown forces are then obtained by applying the equations of equilibrium, i.e.

$$\Sigma H = 0 \tag{7.3}$$
$$\Sigma V = 0 \tag{7.4}$$

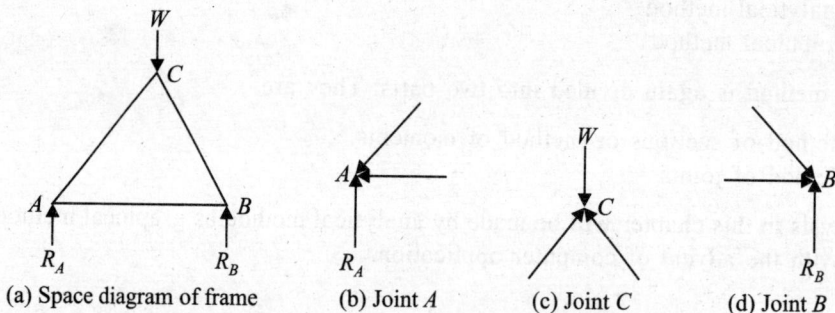

(a) Space diagram of frame　　(b) Joint $A$　　(c) Joint $C$　　(d) Joint $B$

**Figure 7.4**　Determination of forces by method of joint.

**EXAMPLE 7.1**　Find the forces in the members $AB$, $AC$ and $BC$ of the frame shown in Figure 7.5 by method of sections.

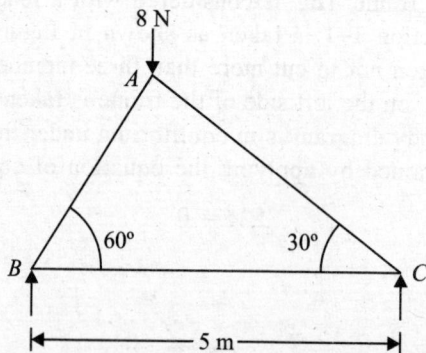

**Figure 7.5**　Visual of Example 7.1.

**Solution**　**Method of sections:**

Find the two reactions $R_B$ and $R_C$ at $B$ and $C$, consider Figure 7.5(a)

From the geometry of the figure,

$$AB = BC \sin 30° = 5 \times 0.5 = 2.5 \text{ m}$$

and

$$BD = AB \sin 30° = 2.5 \times 0.5 = 1.25 \text{ m}$$

It shows that the perpendicular distance of load 8 N from $B$ is at a distance of 1.25 m. Taking moments of the forces about $B$,

$$8 \times 1.25 = R_C \times 5$$

$$\therefore R_C = 2 \text{ N} \qquad \therefore R_B = (8 - 2) = 6 \text{ N}$$

Now consider Figure 7.5(b) in which section 1-1 cuts the sides *BA* and *BC*. Let the directions of forces *AB* and *BC* be taken as shown in Figure 7.5(b).

Taking moment about the point *C*,

$$AB \times 5 \sin 60° = 6 \times 5$$

or $\qquad AB = \dfrac{6 \times 5}{5 \sin 60°} = \dfrac{30}{5 \times 0.866025} = 6.9282 \text{ N (compression)} \qquad \text{(Answer)}$

Again taking moment about the point *A*,

$$BC \times 1.25 \tan 60° = 6 \times 1.25$$

or $\qquad BC = \dfrac{6 \times 1.25}{1.25 \tan 60°} = \dfrac{7.5}{1.25 \times 1.73205} = 3.4641 \text{ N (tension)} \qquad \text{(Answer)}$

Now consider section 2-2 in Figure 7.5(c),

Taking moment about *B*,

$$AC \times 5 \sin 30° = 2 \times 5$$

or $\qquad AC = \dfrac{2 \times 5}{5 \sin 30°} = \dfrac{10}{5 \times 0.5} = 4 \text{ N (compression)} \qquad \text{(Answer)}$

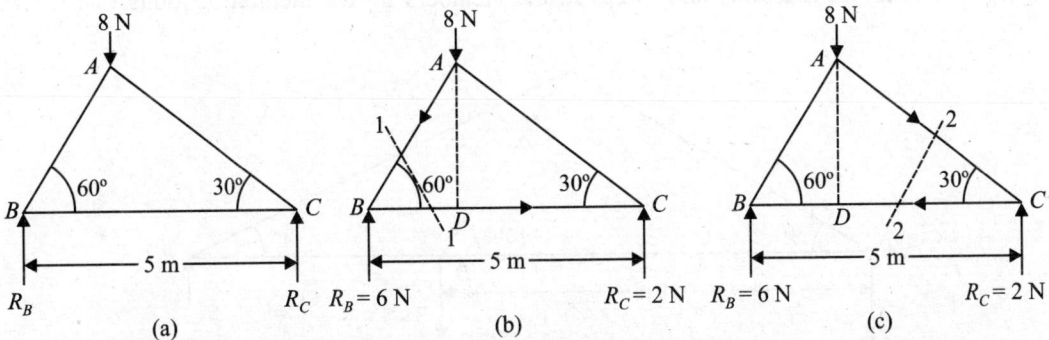

**Figure 7.5**   (a) Find the reaction. (b) First section 1–1. (c) Second section 2–2.

**EXAMPLE 7.2**   Solve Example 7.1 by method of joints.

**Solution**   Consider first joint *B* as shown in Figure 7.6(a). Let the directions of *BC* and *AB* be assumed as shown. Since vertical component of *BC* is zero, direction of *AB* will be downwards.

Resolving vertically the forces at *B*,

$$AB \sin 60° = 6$$

$$AB = \frac{6}{\sin 60°} = 6.92824 \text{ N (compression)} \qquad \text{(Answer)}$$

Resolving forces horizontally at $B$, $AB \cos 60° = BC$

$\therefore$ $\qquad$ $BC = 6.9282 \times \cos 60° = 6.9282 \times 0.5 = 3.4641$ N (tension) $\qquad$ (Answer)

Next, consider the joint $C$ [Figure 7.6(b)]
Resolving the forces vertically,

$$AC \sin 30° = 2 \quad \therefore AC = \frac{2}{\sin 30°} = \frac{2}{0.5} 4 \text{ N (compression)} \qquad \text{(Answer)}$$

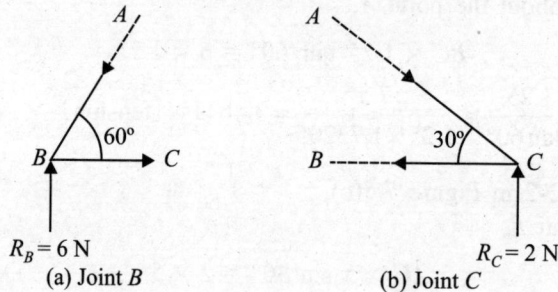

$R_B = 6$ N $\qquad\qquad\qquad\qquad\qquad$ $R_C = 2$ N
(a) Joint $B$ $\qquad\qquad\qquad\qquad\qquad$ (b) Joint $C$

**Figure 7.6** Method of joints.

**EXAMPLE 7.3** A truss of 15 m span carries a point load 2 kN at joint $d$ as shown in Figure 7.7. Find the reactions and forces in the members by the method of joints.

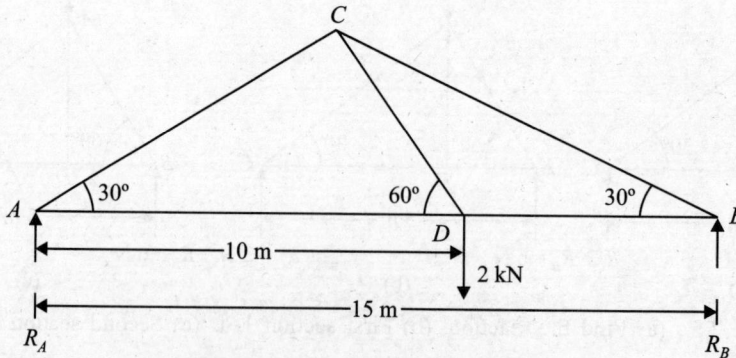

**Figure 7.7** Sketch of Example 7.3.

**Solution** Let us first determine the reactions at $A$ and $B$.
Taking moment about $A$,

$$R_B \times 15 = 2 \times 10$$

or

$$R_B = \frac{2 \times 10}{15} = 1.333 \text{ kN}$$

Hence

$$R_A = 2 - 1.333 = 0.667 \text{ kN}$$

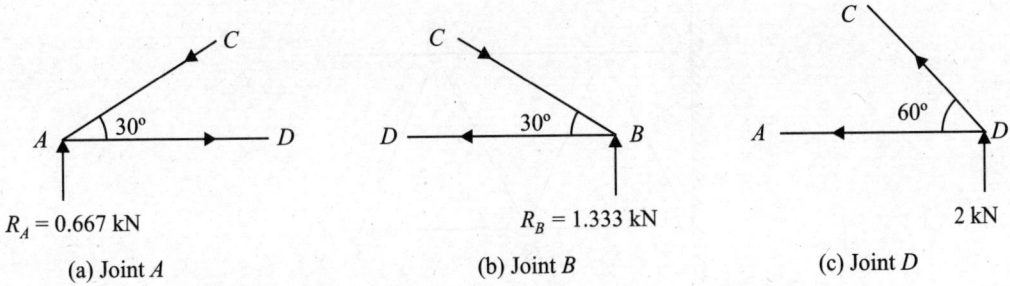

**Figure 7.8**  Solution of Example 7.3 by method of joints.

Consider joint $A$ in the Figure 7.8,
Resolving the forces vertically,

$$AC \sin 30° = R_A$$

$$\text{Force in } AC = \frac{R_A}{\sin 30°} = \frac{0.667}{0.5} = 1.334 \text{ kN (compression)} \qquad \text{(Answer)}$$

Resolving the forces horizontally, force in $AD$,

$$AD = AC \cos 30°$$

$$\therefore \qquad AD = 1.334 \times 0.866025 = 1.1552 \text{ kN (tenslie)}$$

Consider joint $B$,
Resolving forces vertically, forces along $BC$,

$$BC \sin 30° = 1.333$$

$$BC = \frac{1.333}{\sin 30°} = \frac{1.333}{0.5} = 2.667 \text{ kN (compression)} \qquad \text{(Answer)}$$

Resolving forces horizontally, force along $BD$,

$$BD = BC \cos 30°$$

$$BD = 2.667 \times \cos 30° = 2.667 \times 0.866025 = 2.30969 \qquad \text{(Answer)}$$

Considering joint $D$, resolving the forces vertically,

$$DC \sin 60° = 2$$

$$DC = \frac{2}{\sin 60°} = \frac{2}{0.866025} = 2.3094 \text{ kN (tension)} \qquad \text{(Answer)}$$

## 7.7  CANTILEVER TRUSSES

A truss or frame connected to a wall or a column at one end with other end free is shown in
Figure 7.9. Such truss is called *cantilever truss*. When this cantilever truss is loaded,
the determination of support reactions is not essential. The determination of forces in the
different members can be started from the free end.

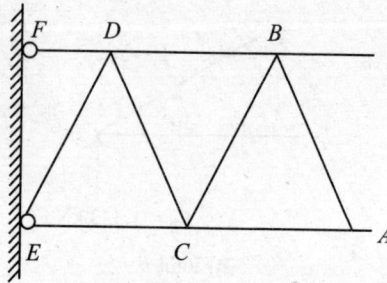

**Figure 7.9** Cantilever truss or frame.

**EXAMPLE 7.4** Determine the forces in all the members of the cantilever truss shown in Figure 7.10.

**Solution** Consider Figure 7.10. Forces in the different members can be solved by method of joints or methods of sections.

Take method of joints. Consider joint $C$. The forces acting on $C$ are shown in Figure 7.11(a).

Resolving the forces vertically,

$$CD \sin 60° = 2000$$

or $$CD = \frac{2000}{\sin 60°} = 2309.401 \text{ N (tension)} \quad \text{(Answer)}$$

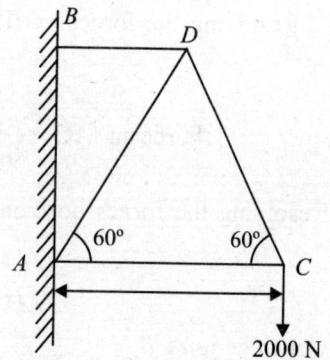

**Figure 7.10** Sketch of Example 7.4.

Resolving the forces horizontally,

$$AC = CD \cos 60° = 2309.401 \times 0.5 = 1154.7005 \text{ N (compression)} \quad \text{(Answer)}$$

Considering the forces at joint in Figure 7.11(b).

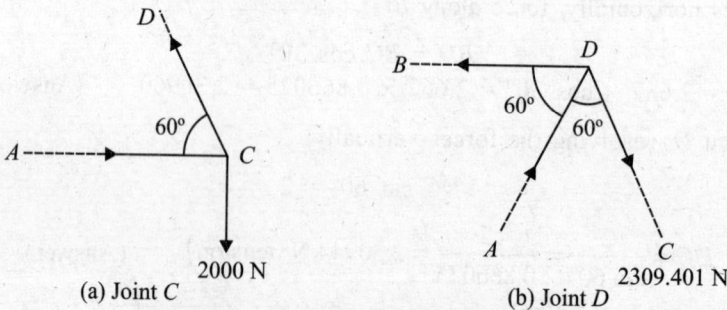

(a) Joint $C$  (b) Joint $D$

**Figure 7.11** Method of joints at $C$ and at $D$.

Resolving the forces vertically,

$$AD \sin 60° = 2309.401 \times \sin 60°$$

or $$AD = 2309.401 \text{ N (compression)} \quad \text{(Answer)}$$

Resolving the forces horizontally,

$$DB = 2309.401 \cos 60° + AD \cos 60°$$
$$DB = 2309.401 \times 0.5 + 2309.401 \times 0.5 = 2309.401 \text{ (tension)}  \text{(Answer)}$$

## 7.8  CONCLUSION

The chapter is confined to the analysis of perfect frame only. Therefore, difference of perfect and imperfect frames are explained at the very beginning. Imperfect frame is again classified into deficient and redundant frame analysis which is outside the scope of this chapter. Two different methods of analysis of forces in perfect frame under equilibrium loading conditions are presented. Both the methods of solution, i.e. method of joints and method of sections are quite good and equally easy. Cantilever frame is also presented. A few numerical examples are solved by both the methods.

## EXERCISES

**7.1** Find the forces in members $AB$, $BC$ and $AC$ of the frame, as shown in Figure 7.12, by any one of the methods you like.

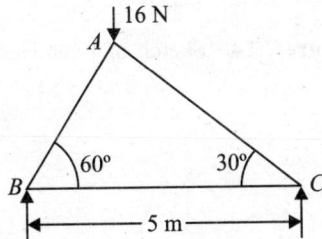

**Figure 7.12**   Sketch of Exercise 7.1.

(Answer    Force on $AB$ = 13.8564 N (compression)
Force on $AC$ = 8 N (compression)
Force on $BC$ = 6.9282 N (tension))

**7.2** A truss of span 7.5 m is loaded as shown in Figure 7.13. Find the forces in the members of the truss.

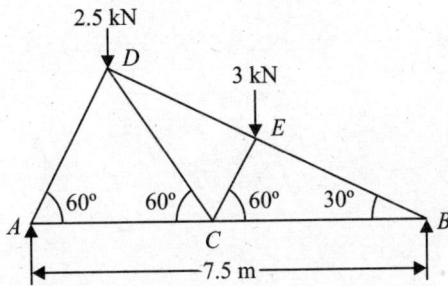

**Figure 7.13**   Sketch of Exercise 7.2.

(Answer    $AD = 3.464$ kN (compression)
$AC = 1.732$ kN (tension)
$CD = 2.598$ kN (tension)
$CE = 2.598$ kN (compression)
$DE = 3.5$ kN (compression)
$BE = 5$ kN (compression)
$BC = 4.33$ kN (tension))

**7.3** A cantilever truss of Warren type is loaded as shown in Figure 7.14. Find forces in members $BC$, $BE$, $BD$ and $AB$ of the truss.

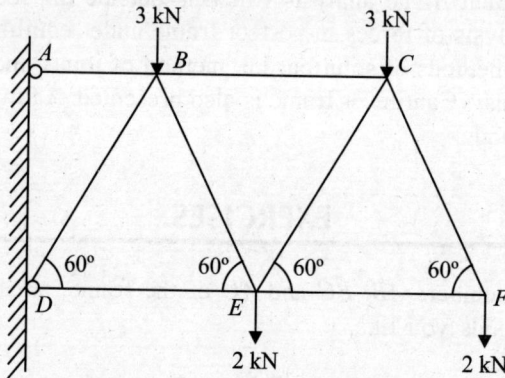

**Figure 7.14**    Sketch of Exercise 7.3.

(Answer    $BC = 4.04$ N
$BE = 8.08$ N
$BD = 11.54$ N
$AB = 11.85$ N)

# CHAPTER 8

# Centre of Gravity

## 8.1 INTRODUCTION

It is known that force of gravity of the earth attracts every particle towards its centre. This force of attraction which is always proportional to the mass of the body is called *weight*. The point through which the weight of the body acts towards the centre is called *centre of gravity* (CG) or simply *G* as shown in Figure 8.1.

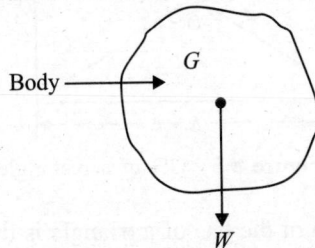

**Figure 8.1**  The weight *W* of the body acts through *G*.

Again, the plane figures like triangles, quadrilateral, circle, ellipse, etc. have only areas but no mass. Centre of area of such figures is called *centroid*. The methods of finding both centre of gravity and centroid are similar. It is quite often the centroid is also known as centre of gravity *G*.

## 8.2 METHOD OF DETERMINING THE CENTRE OF GRAVITY (CG)

The following are the methods to determine the centre of gravity.

(i) Geometrical consideration
(ii) Method of moments
(iii) Method of graphics

## 8.2.1   Geometrical Consideration

The following geometrical figures give the CG

(i) **Uniform rod:**   The CG of a uniform rod is always at the middle as shown in Figure 8.2

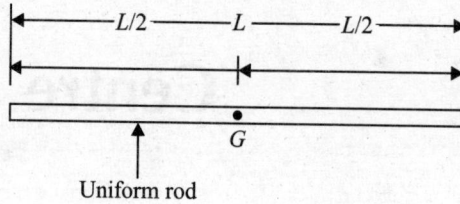

Uniform rod

**Figure 8.2**   CG of uniform rod.

(ii) **Rectangle:**   Figure 8.3 shows the position of the CG of a rectangle (or parallelogram) of length $L$ and breadth $B$. It is the point of intersection of the two diagonals.

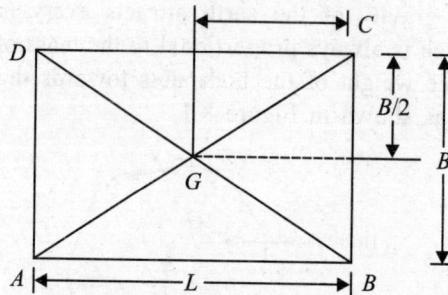

**Figure 8.3**   CG of a rectangle.

(iii) **Triangle:**   The position of the CG of a triangle is the points where the three medians intersect at a point as shown in Figure 8.4. Medians are the line joining the vertices and its opposite sides at the middle points.

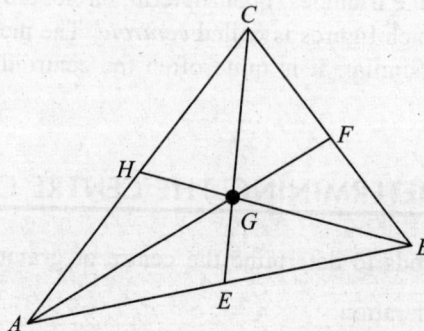

**Figure 8.4**   CG of the triangle.

**(iv) Semi-circle:** The CG of a semicircle is at a distance of $\dfrac{4r}{3\pi}$ from the base $AB$ along the vertical radius $OC$ as shown in Figure 8.5(a).

**(v) Trapezium:** If $a$ and $b$ are the two parallel sides of a trapezium and $h$ is the distance between them as shown in Figure 8.5(b), the distance of the CG from the longer parallel side of length $b$ is $\dfrac{h}{3}\dfrac{(b+2a)}{(b+a)}$

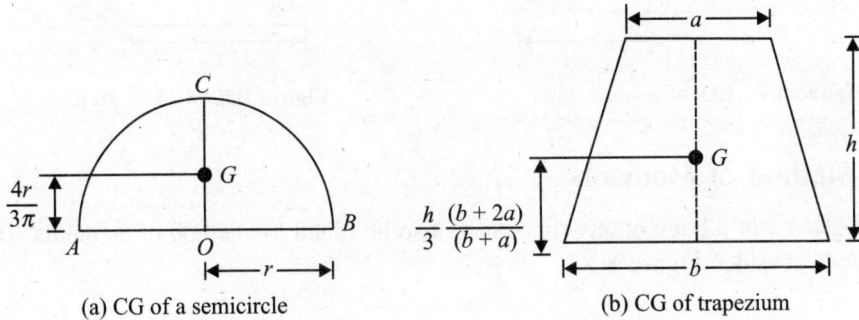

(a) CG of a semicircle                    (b) CG of trapezium

**Figure 8.5**   CG of semicircle and trapezium.

**(vi) Right-circular solid cone:** If $h$ is the height of the right-circular solid cone, its CG is at a height of $\dfrac{h}{4}$ from the base as shown in Figure 8.6.

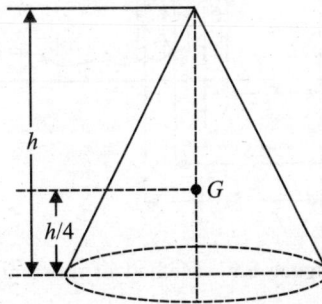

**Figure 8.6**   CG of a right-circular solid cone.

**(vii) Hemisphere:** If $r$ is the radius of the hemisphere, its CG lies at a distance $\dfrac{3r}{8}$ from the base as shown in Figure 8.7.

**(viii) Cube:** The CG of the cube lies at a distance of $\dfrac{l}{2}$ from every face. Here $l$ is the length of each side (Figure 8.8).

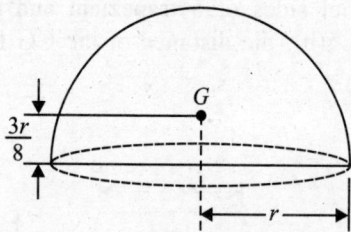

**Figure 8.7** CG of a hemisphere.

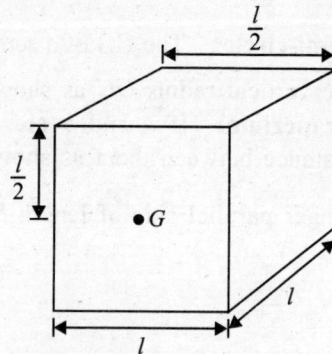

**Figure 8.8** CG of a cube.

## 8.2.2 Method of Moments

Centre of gravity of a body of any shape may also be found by method of moments. To explain this method, consider Figure 8.9.

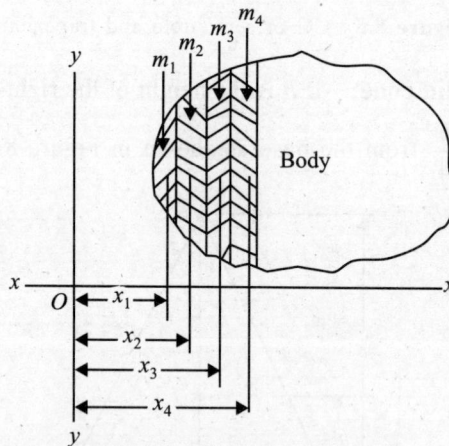

**Figure 8.9** Method of moments.

Let the whole body be the mass $M$ whose CG is to be determined. Divide the whole mass in a number of small masses of particles $m_1$, $m_2$, $m_3$, $m_4$, ... and let $(x_1, y_1)$, $(x_2, y_2)$, $(x_3, y_3)$ $(x_4, y_4)$, ... be their corresponding coordinates of centres of gravity with respect to coordinate system $xox$ and $yoy$ in Figure 8.9.

Let $\bar{x}$ and $\bar{y}$ be the coordinates of the CG of the body. Then from principle of moments,

$$M\,\bar{x} = m_1 x_1 + m_2 x_2 + m_3 x_3 + m_4 x_4 \ldots = \Sigma mx$$

$\therefore$
$$\bar{x} = \frac{\Sigma mx}{M} \tag{8.1}$$

Similarly

$$\bar{y} = \frac{\Sigma my}{M} \qquad (8.2)$$

Here
$$M = m_1 + m_2 + m_3 + m_4 \qquad (8.3)$$

## 8.2.3  Centre of Gravity of Plane Figures (Method of Graphics)

The plane geometrical figures such as L-section, T-section, I-section, etc. have only areas. The CG of such bodies is obtained in the same procedures as that of solid body.

Let $\bar{x}$ and $\bar{y}$ be the coordinates of the CG with respect to some coordinate axes. If $a_1, a_2, a_3, a_4...$ are the areas into which the figure is divided, $x_1, x_2, x_3, x_4...$ are the respective coordinates on $xox$ and $y_1, y_2, y_3, y_4, ...$ are in $yoy$ axis, then,

$$\bar{x} = \frac{a_1 x_1 + a_2 x_2 + a_3 x_3 + a_4 x_4 + \cdots}{a_1 + a_2 + a_3 + a_4 + \cdots} = \frac{\sum a_i x_i}{A} \qquad (8.4)$$

Similarly

$$\bar{y} = \frac{a_1 y_1 + a_2 y_2 + a_3 y_3 + a_4 y_4 + \cdots}{a_1 + a_2 + a_3 + a_4 + \cdots} = \frac{\sum a_i y_i}{A} \qquad (8.5)$$

In calculating $x_1, x_2, x_3, x_4, ...$ and $y_1, y_2, y_3, y_4, ...$ **axis of reference** is to be selected for the plane figures. Generally for $x_1, x_2, x_3, x_4,...$ left line of figure and for $y_1, y_2, y_3, y_4, ...$ bottom or lower line are taken.

Again if the figure is symmetrical about $xox$ or $yoy$ axis, the CG of the body lies in the **axis of symmetry**. In such cases, either $\bar{x}$ and $\bar{y}$ is to be calculated for the CG.

*EXAMPLE 8.1*  Find the CG of the T-section shown in Figure 8.10.

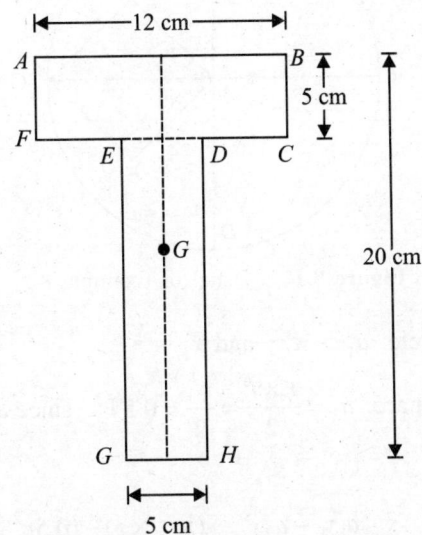

**Figure 8.10**  Visual of Example 8.1.

*Solution*    The given T-section is symmetrical about $y$-axis shown by dotted line and, therefore, CG will lie on this line. Let $G$ be the centre of gravity of the T-section and $\bar{y}$ be the distance $G$ from the bottom line $GH$.

Now area $ABCF = a_1 = 12 \times 5 = 60$ cm²

and $$y_1 = (20 - 2.5) = 17.5 \text{ cm}$$

Again area $DHGE = a_2 = (20 - 5) \times 5 = 75$ cm²

and $$y_2 = \frac{20-5}{2} = 7.5 \text{ cm}$$

Applying Eq. (8.5)

$$\bar{y} = \frac{a_1 y_1 + a_2 y_2}{a_1 + a_2} = \frac{60 \times 17.5 + 75 \times 7.5}{60 + 75} = 11.9444 \text{ cm} \qquad \text{(Answer)}$$

**EXAMPLE 8.2**    A square hole is punched out of circular lamina, the diagonal of the square being a radius of the circle. Find the CG of the remainder and also show that this new CG is at a distance of $\dfrac{2r}{8\pi - 4}$ from the centre of the circle.

*Solution*    The section in Figure 8.11 is symmetrical about $x$-axis. Hence its CG will lie in $x$-axis. Let $G_n$ be its new CG. Let $x$ be the distance between the point $A$ and new centre of gravity $G_n$.

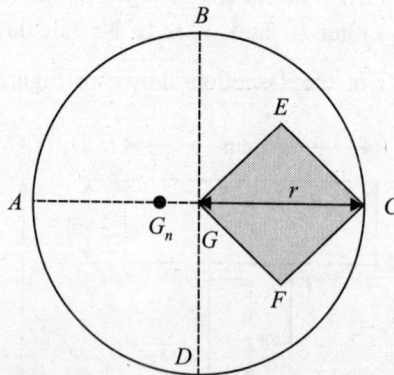

**Figure 8.11**    Visual of Example 8.2.

Considering the main circle, $a_1 = \pi r^2$ and $x_1 = r$

Considering the cut out square, $a_2 = \dfrac{r \times r}{2} = \dfrac{r^2}{2} = 0.5\,r^2$, since area of a square = (diagonal × diagonal)/2.

Using the equation

$$x = \frac{a_1 x_1 - a_2 x_2}{a_1 - a_2} = \frac{(\pi r^2 \times r) - (0.5 r^2 \times 1.5 r)}{\pi r^2 - 0.5 r^2}$$

Simplifying

$$x = \frac{\pi r - 0.75r}{\pi - 0.5} \qquad \text{(Answer)}$$

Distance between the $G_n$ and $G$ is:

$$(r - \overline{x}) = r - \frac{\pi - 0.75}{\pi - 0.5} = \frac{\pi r - 0.5r = \pi r + 0.75r}{\pi - 0.5}$$

$$\overline{x} = \frac{0.25r}{\pi - 0.5} = \frac{2r}{8\pi - 4} \qquad \text{Shown}$$

**EXAMPLE 8.3**   Find the centre of gravity of an L-section shown in Figure 8.12.

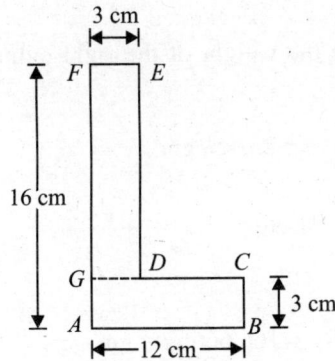

**Figure 8.12**   Visual of Example 8.3.

**Solution**   The given L-section is not symmetrical. Therefore, two axes of reference will have to be considered to find the CG of the section. Here is to be calculated with reference to the bottom line $AB$ and $\overline{x}$ is to be calculated with reference to the left line $AF$.

Divide the L-section into rectangles *FEDG* and *ABCG*.

Let $\qquad\qquad a_1$ = area of rectangle *FEDG* = $(16 - 3) \times 3 = 39$ cm$^2$
and $\qquad\qquad y_1 = 3 + (16 - 3)/2 = 9.5$ cm

Again $a_2$ = area of rectangle *ABCG* = $12 \times 3 = 36$ cm$^2$

and $\qquad\qquad\qquad y_2 = 3/2 = 1.5$ cm
Using Eq. (8.5)

$$\overline{y} = \frac{a_1 y_1 + a_2 y_2}{a_1 + a_2} = \frac{39 \times 9.5 + 36 \times 1.5}{39 + 36} = 5.66 \text{ cm}$$

Again

$$x_1 = 3/2 = 1.5 \text{ cm}$$

and

$$x_2 = 12/2 = 6 \text{ cm}$$

Using Eq. (8.4)

$$\bar{x} = \frac{a_1 x_1 + a_2 x_2}{a_1 + a_2} = \frac{39 \times 1.5 + 36 \times 6}{39 + 36} = 3.66 \text{ cm}$$

Hence CG of the L-section is at a distance of 5.66 cm from the bottomline $AB$ and 3.66 cm from the left line $AF$.     (Answer)

**EXAMPLE 8.4**   A right circular cylinder of 10 cm diameters joined with a hemisphere of the same diameter face to face. Find the greatest height of the cylinder so that CG of the composite section coincides with the plane of joining the two sections. The density of the material of the hemispherical part is twice the density of the material of the cylinder.

**Solution**   The joined cylinder and the hemisphere shown in Figure 8.13 is symmetrical about vertical axis $AOB$.

Considering the right cylinder,

Let $h$ be the height and $w_1$ be the weight of the right cylinder. Then

$$w_1 = \rho \times \frac{\pi}{4} \times (10)^2 h = 25\rho\pi h \text{ gm}$$

and

$$y_1 = 5 + \frac{h}{2} = 5 + 0.5h \text{ cm}$$

Considering the hemisphere,

$$w_2 = 2\rho \times \frac{2\pi}{3} \times (5)^3 = 166.6666 \, \rho\pi \text{ gm}$$

and

$$y_2 = 5 - \frac{3r}{8} = 5 - \frac{3 \times 5}{8} = \frac{25}{8} \text{ cm}$$

Using the equation

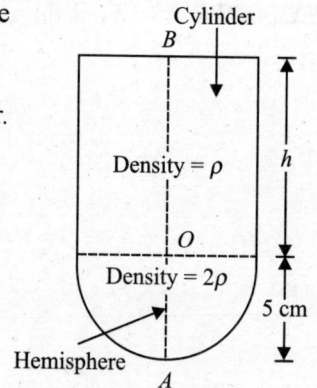

**Figure 8.13**   Visual of Example 8.4.

$$\bar{y} = \frac{w_1 y_1 + w_2 y_2}{w_1 + w_2} = \frac{25\rho\pi h \times (5 + 0.5h) + 166.6666\rho\pi \times \dfrac{25}{8}}{25\rho\pi h + 166.6666\rho\pi}$$

or

$$5 = \frac{25h \times (5 + 0.5h) + 520.833}{25h + 166.6666}$$

or   $125h + 833.3345 = 125h + 12.5h2 + 520.833$

or   $\qquad 12.5h^2 = 833.3345 - 520.833 = 312.5012$

or

$$h = \sqrt{\frac{312.5012}{12.5}} = \sqrt{25} = 5 \text{ cm} \qquad \text{(Answer)}$$

**EXAMPLE 8.5**   Determine the CG of the area, shown in Figure 8.14, by taking moment of area about the given x-x axis and y-y axis.

**Solution**   The given area is divided into three parts as shown in Figure 8.14 as:

Area $ABCD = 1$, area $BEF = 2$, area $CDE = 3$

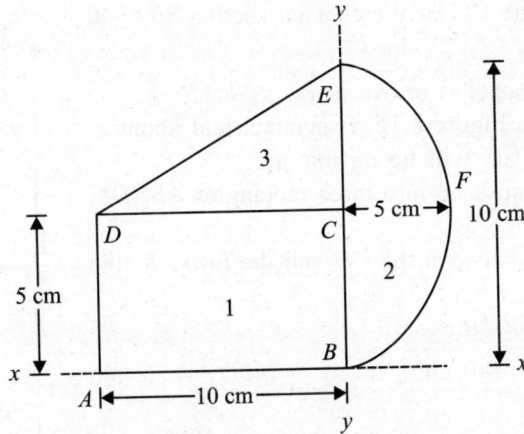

**Figure 8.14**  Visual of Example 8.5.

Let $\bar{x}$ be the distance of centroid from $x$-$x$

and $\bar{y}$ be the distance of centroid from $y$-$y$

Now area

$A_1 = 10 \times 5 = 50 \text{ cm}^2$,     $A_2 = \dfrac{\pi r^2}{2} = \dfrac{\pi \times 5^2}{2} = 12.5\pi \text{ cm}^2$,     $A_3 = \dfrac{1}{2} \times 10 \times 5 = 25 \text{ cm}^2$

Corresponding

$x_1 = 5 \text{ cm}$,     $x_2 = \dfrac{4r}{3\pi} = \dfrac{4 \times 5}{3\pi} = 2.122 \text{ cm}$,     $x_3 = 10/3 = 3.333 \text{ cm}$,

and

$y_1 = 2.5 \text{ cm}$,     $y_2 = 5 \text{ cm}$,     $y_3 = 5 + 5/3 = 6.666 \text{ cm}$

Now use equation

$$\bar{x} = \frac{A_1 x_1 + A_2 x_2 + A_3 x_3}{A_1 + A_2 + A_3} = \frac{50 \times (-5) + 12\pi \times 2.122 + 25 \times (-3.333)}{50 + 12\pi + 25}$$

$\therefore$     $$\bar{x} = \frac{-250 + 79.9975 - 83.325}{112.6991} = \frac{-253.3275}{112.6991} = -2.2478 \text{ cm}$$

(–ve sign indicates that the centroid is left axis $y$-$y$).

Again use the equation

$$\bar{y} = \frac{A_1 y_1 + A_2 y_2 + A_3 y_3}{A_1 + A_2 + A_3} = \frac{50 \times 2.5 + 12\pi \times 5 + 25 \times 6.666}{50 + 12\pi + 25}$$

or     $$\bar{y} = \frac{125 + 188.4855 + 166.65}{112.6991} = 4.26 \text{ cm}$$

(+ve sign indicates that the centroid is above the line $x$-$x$)

The centroid is (–2.2478 cm, 4.26 cm)     (Answer)

**EXAMPLE 8.6** Find the CG of the channel section $80 \times 40 \times 12$ mm.

**Solution** The given channel is drawn in Figure 8.15.

The section, shown in Figure 8.15, is symmetrical about $x$-axis and, therefore, the CG will lie on this axis.

The whole section is spilt up into three rectangles *ABCDE*, *DEFG* and *FGHJK*.

Let $\bar{x}$ be the distance between the CG and the face *AK*, the axis of reference.

Consider rectangle *ABCDE*,

$$a_1 = 40 \times 12 = 480 \text{ mm}^2, \text{ and } x_1 = (40/2) = 20 \text{ mm}$$

For rectangle *DEFG*,

$$a_2 = (80 - 24) \times 12 = 672 \text{ mm}^2, \text{ and } x_2 = 12/2 = 6 \text{ mm}$$

For rectangle *FGHJK*,

$$a_3 = 40 \times 12 = 480 \text{ mm}^2, \text{ and } x_3 = (40/2) = 20 \text{ mm}$$

Using equation

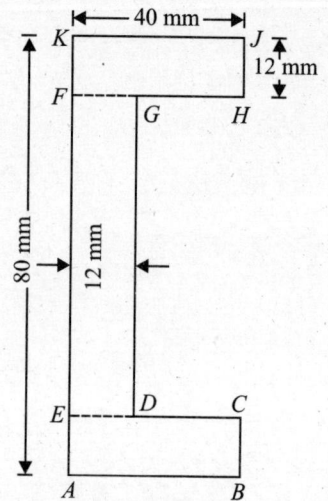

**Figure 8.15** Visual of Example 8.8.

$$\bar{x} = \frac{a_1 x_1 + a_2 x_2 + a_3 x_3}{a_1 + a_2 + a_3}$$

$$\bar{x} = \frac{(480 \times 20) + (672 \times 6) + (480 \times 20)}{480 + 672 + 480}$$

or

$$\bar{x} = \frac{23232}{1632} = 14.2353 \text{ mm} \qquad \text{(Answer)}$$

**EXAMPLE 8.7** An I-section is made of two flanges of different sizes and a web as shown in Figure 8.16. Find height of CG from the bottom flange.

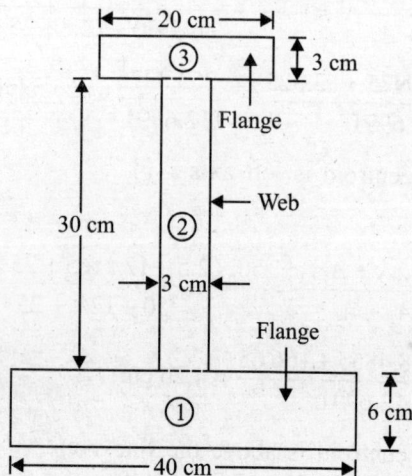

**Figure 8.16** Visual of Example 8.7.

*Solution*    The section, shown in Figure 8.16, is symmetrical about the y-axis. Hence CG of the section will lie on this axis. To find the height $\bar{y}$ of CG from the bottom flange, divide the whole section into three rectangles 1, 2 and 3 as shown.

Considering rectangle 1,

$$a_1 = 40 \times 6 = 240 \text{ cm}^2 \text{ and } y_1 = (6/2) = 3 \text{ cm}$$

From rectangle 2,

$$a_2 = 30 \times 3 = 90 \text{ cm}^2 \text{ and } y_2 = 6 + (30/2) = 21 \text{ cm}$$

From rectangle 3,

$$a_3 = 20 \times 3 = 60 \text{ cm}^2 \text{ and } y_3 = 6 + 30 + (3/2) = 37.5 \text{ cm}$$

Using the equation

$$\bar{y} = \frac{a_1 y_1 + a_2 y_2 + a_3 y_3}{a_1 + a_2 + a_3}$$

$$\bar{y} = \frac{(240 \times 3) + (90 \times 21) + (60 \times 37.5)}{240 + 90 + 60}$$

$$\bar{y} = \frac{4860}{390} = 12.4615 \text{ cm} \qquad \text{(Answer)}$$

**EXAMPLE 8.8**    A body consists of right circular solid cone of height 100 cm and radius of 80 cm placed on solid hemisphere of radius equal to 80 cm radius of same material. Find the position of the CG of body.

*Solution*    The sketch of Example 8.8 with its right circular cone and solid hemisphere is drawn in Figure 8.17. Thus the body, shown in Figure 8.17, is symmetrical about y-axis in which the CG will lie. Let $\bar{y}$ be the distance between the CG and D, the point of reference.

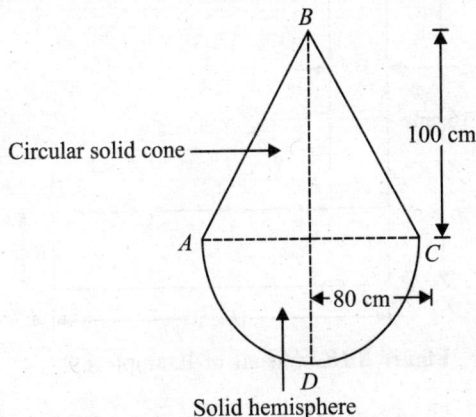

**Figure 8.17**    Visual of Example 8.8.

We know the formula of the volume of right circular cone $v_1 = \dfrac{\pi}{3} \times r^2 \times h$, where r is the radius at base and h is the height.

Hence

$$\text{volume} = v_1 = \frac{\pi}{3} \times r^2 \times h = \frac{\pi}{3} \times 80^2 \times 100 = 670206.4328 \text{ cm}^2$$

and

$$y_1 = 80 + \frac{100}{4} = 105 \text{ cm} \qquad \left( \because \text{CG of right circular cone} = \frac{h}{4} \right)$$

We know the formula of volume of hemisphere $= v_2 = \dfrac{1}{2} \times$ volume of sphere

or

$$v_2 = \frac{1}{2} \times \frac{4}{3} \times \pi \times r^3 = \frac{2}{3} \times \pi \times 80^3 = 1072330.292 \text{ cm}^3$$

and

$$y_2 = r - \frac{3}{8}r = r\left(1 - \frac{3}{8}\right) = \frac{5}{8}r = \frac{5}{8} \times 80 = 50 \text{ cm}$$

Now use the equation

$$\overline{y} = \frac{v_1 y_1 + v_2 y_2}{v_1 + v_2} = \frac{670206.4328 \times 105 + 1072330.292 \times 50}{670206.4328 + 1072330.292}$$

$$\therefore \qquad \overline{y} = 71.1538 \text{ cm} \qquad \text{(Answer)}$$

**EXAMPLE 8.9**   Find the CG of the Z-section shown in Figure 8.18.

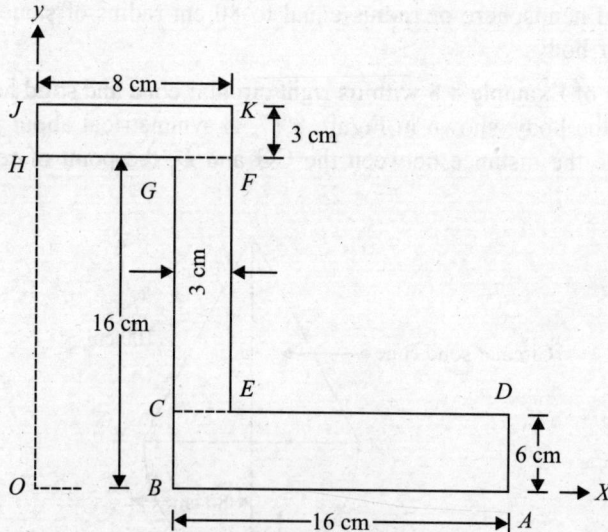

**Figure 8.18**   Visual of Example 8.9.

**Solution**   Z-section of Example 8.9 is shown in Figure 8.18.

The section is not symmetrical about x-axis or y-axis and, therefore, both $\overline{x}$ and $\overline{y}$ are to be calculated with *OA* as x-axis and *OJH* as y-axis.

Divide the Z-section into three rectangles as *ABCED*, *CEFG* and *FGHJK*.
Considering rectangles *ABCED* with reference line *OHJy*,

$$a_1 = 16 \times 6 = 96 \text{ cm}^2 \text{ and } x_1 = (8 - 3) + (16/2) = 13 \text{ cm}, \ y_1 = (6/2) = 3 \text{ cm}$$

Similarly for rectangle *CEFG*,

$a_2 = 10 \times 3 = 30$ cm$^2$ and $x_2 = (8 - 3) + (3/2) = 6.5$ cm, $y_2 = 6 + (10/2) = 11$ cm

Similarly, for rectangles *FGHJK*

$a_3 = 8 \times 3 = 24$ cm$^2$ and $x_3 = (8/2) = 4$ cm, $y_3 = 16 + (3/2) = 17.5$ cm

Using equation

$$\bar{x} = \frac{a_1x_1 + a_2x_2 + a_3x_3}{a_1 + a_2 + a_3} = \frac{96 \times 13 + 30 \times 6.5 + 24 \times 4}{96 + 30 + 24} = 10.26 \text{ cm} \qquad \text{(Answer)}$$

Using equation

$$\bar{y} = \frac{a_1y_1 + a_2y_2 + a_3y_3}{a_1 + a_2 + a_3} = \frac{96 \times 3 + 30 \times 11 + 24 \times 17.5}{96 + 30 + 24} = 6.92 \text{ cm} \qquad \text{(Answer)}$$

## 8.3   CONCLUSION

The definitions of centre of gravity and methods to determine it have been given. Centre of gravities of triangle, rectangle, uniform rod, trapezium, semicircle, right circular solid cone, hemisphere and cube are given. The method of moments has been described. The most important equations of finding the centre of gravity of plane figures are deduced and presented in Eqs. (8.4) and (8.5). Various numerical examples of finding the centre of gravity of T-section, L-section, Z-section, I-section, composite section, etc. are solved.

## EXERCISES

**8.1** Define centre of gravity (CG). Distinguish between centre of gravity and centroid.
**8.2** Describe the method of finding the centre of gravity of a body.
**8.3** Where does the CG of the following section lie?
(a) Semi-circle                 (b) Hemisphere
(c) Trapezium               (d) Right circular solid cone
**8.4** Find the centre of gravity (CG) of the T-section shown in Figure 8.19.

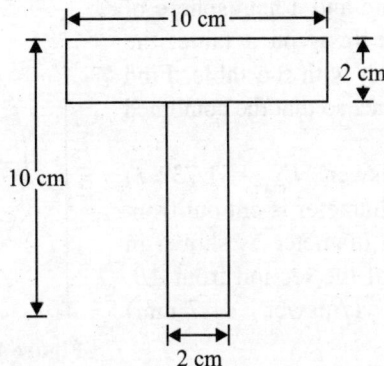

**Figure 8.19**   Visual of Exercise 8.4.   (Answer   $\bar{x}$ = 8.555 cm)

**8.5** Find the centre of gravity (CG) of the I-section shown in Figure 8.20.

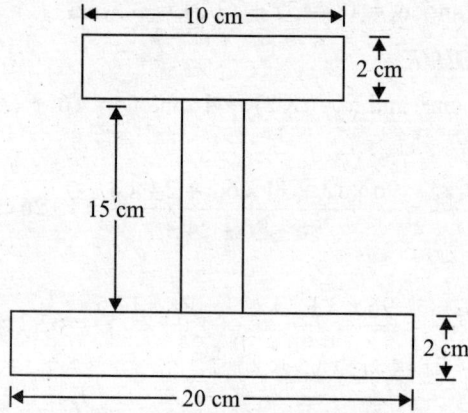

**Figure 8.20**  Visual of Exercise 8.5.    (Answer  $\bar{x}$ = 7.6111 cm)

**8.6** Find the centre of gravity (CG) of the L-section shown in Figure 8.21.

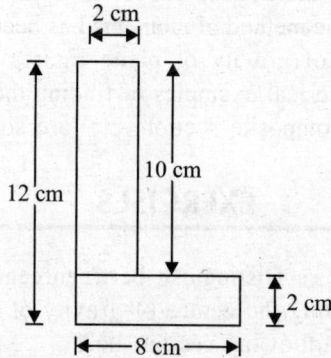

**Figure 8.21**  Visual of Exercise 8.6.

(Answer  $\bar{x}$ = 2.333 cm, $\bar{y}$ = 4.8333 cm)

**8.7** A body consisting of a cone and a hemisphere of radius $r$ on the same base rests on a table, the hemisphere being in contact with the table. Find the greatest height of the cone, so that the combined solid may stand upright.

(Answer  $h_{max}$ = 1.732 $r$)

**8.8** A circular disc of 50 mm diameter is cut out from a circular disc of 100 mm diameter as shown in Figure 8.22. Find the CG of the section from $AB$.

(Answer  41.7 mm)

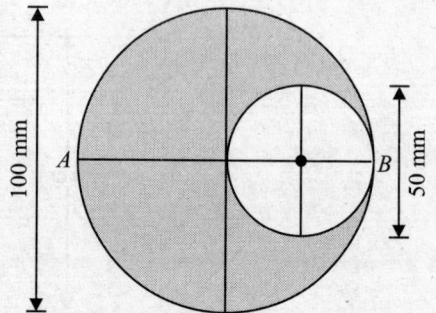

**Figure 8.22**  Visual of Exercise 8.8.

# CHAPTER 9

# Moment of Inertia

## 9.1 INTRODUCTION

The moment ($M$) of a force is the product of the force ($F$) multiplied by the perpendicular distance ($x$) between the point and the line of action of the force and this moment $M = Fx$ (Figure 9.1). If this moment is again multiplied by the perpendicular distance $x$, then the moment (i.e. $Fx^2$) of a force or second moment of a force is called *moment of inertia* (MI). Similarly, the moment of a lamina of area $A$ about $y$-axis (Figure 9.2) is $Ax$. If this moment of area is again multiplied by $x$, second moment of area $A$ will be equal to $Ax^2$. This second moment of area is also called *moment of inertia of the lamina* about $y$-axis. If instead of area, mass of the body is considered, second moment of mass is also called *moment of inertia of mass*. It is normally represented by $I$.

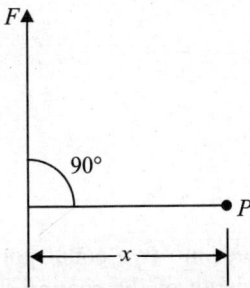

**Figure 9.1** Moment of force $F$.

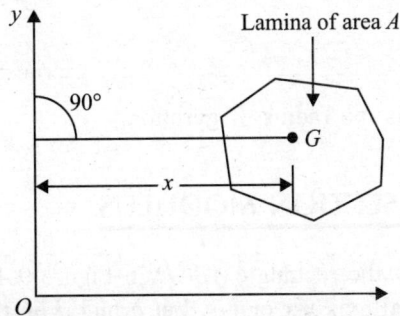

**Figure 9.2** Moment of a lamina of area $A$.

## 9.2 MOMENT OF INERTIA OF A PLANE AREA

In Figure 9.3, the plane area $A$ is divided into a number of small areas like $a_1, a_2, a_3, \dots$ . Let the distances of those divided small areas from a line $y$-$y$, about which moment of inertia $I$ of plane area $A$ is to be obtained, be $x_1, x_2, x_3 \dots$ .

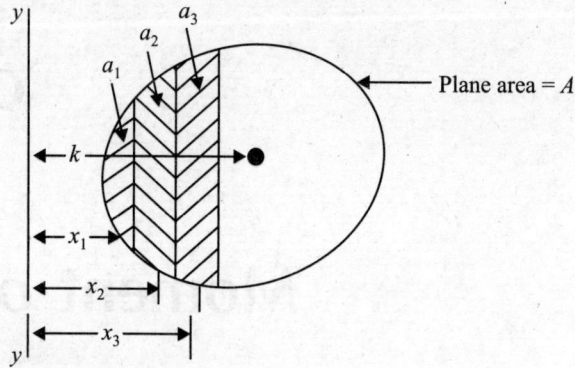

**Figure 9.3**   Moment of inertia of plane area.

Now moment of inertia

$$I = a_1 r_1^2 + a_2 r_2^2 + a_3 r_3^2 + \cdots$$
$$I = \Sigma a r^2 \tag{9.1}$$

## 9.3   RADIUS OF GYRATION

The radius of gyration of a body or plane area about an axis is a distance such that its square multiplied by the area gives the moment of inertia. In Figure 9.3, the moment of inertia of the body about an axis $y$-$y$ is given in Eq. (9.1).

Let the mass or area of the body is concentrated at a distance $k$ from the axis of reference $y$-$y$ as shown in Figure 9.3. Then moment of inertia of the whole mass or area about $y$-$y$ is:

$$I = Ak^2$$

or

$$k = \sqrt{\frac{I}{A}} \tag{9.2}$$

Here $k$ is the radius of gyration.

## 9.4   SECTION MODULUS

Consider the rectangle $ABCD$ in Figure 9.4 which is divided into two equal portions by the centroidal axis $x$-$x$ or $y$-$y$. Let $b$ and $d$ be the breadth and depth of the section. If $I_{xx}$ and $I_{yy}$ are the moments of inertia of the section about $x$-$x$ and $y$-$y$ respectively, then section modulus $Z$ of the section $ABCD$ about $x$-$x$ is:

$$Z_{xx} = \frac{I_{xx}}{\dfrac{d}{2}} . \text{ And about } y\text{-}y \text{ is: } Z_{yy} = \frac{I_{yy}}{\dfrac{b}{2}} \tag{9.3}$$

Thus, Eq. (9.3) gives the section modulus of the section about $x$-$x$ and $y$-$y$ axes.

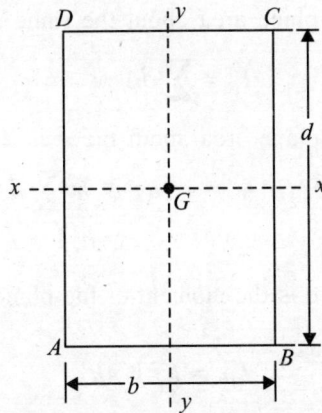

**Figure 9.4** Section modulus.

## 9.5 THEOREM OF PARALLEL AXIS

Theorem of parallel axis states that if the moment of inertia of plane area $A$ about an axis through its centre of gravity $G$, be denoted by $I_G$, the moment of inertia of the area about an axis $AB$ parallel to the first, and at a distance $h$ from the centre of gravity is given by:

$$I_{AB} = I_G + Ah^2 \qquad (9.4)$$

***Proof*** As shown in Figure 9.5, a plane area of $A$ is considered at distance $h$ from the centre of gravity $G$ of the plane area.

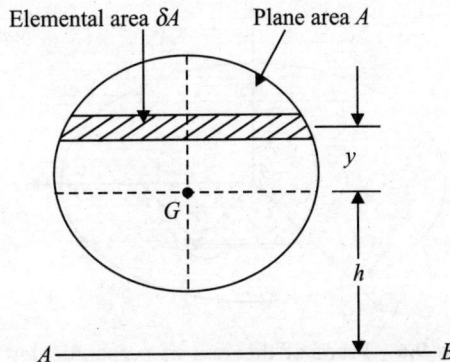

**Figure 9.5** Proof of parallel axis theorem.

A small strip of elemental area $\delta A$ is considered at a distance $h$ from the centre of gravity $G$ of the body.

Moment of inertia of this strip of area $\delta A$ about an axis passing through $G$ at a distance

$$h = \delta A \cdot y^2.$$

Moment of inertia of the whole plane area about the same axis through $G$ is

$$I_G = \sum \delta A \cdot y^2$$

Moment of inertia of the whole plane area about the axis $AB$ parallel to the axis through

$$G = I_{AB} = \sum \delta A (h + y)^2 = \sum \delta A (h^2 + y^2 + 2hy) = h^2 \sum \delta A + \sum \delta A \cdot y^2 + 2h \sum \delta A \cdot y$$
$$= h^2 A + I_G + 0$$

$\because$ The term $\sum \delta Ay$ in the above is the moment of the plane area $A$ about $G$ which is equal to zero.

$\therefore$
$$I_{AB} = I_G + Ah^2$$

## 9.6   THEOREM OF PERPENDICULAR AXIS

It states that if $I_{xx}$ and $I_{yy}$ be the moments of a plane section about two perpendiculars axes meeting at $O$, the moment of inertia $I_{zz}$ about the axis $z$-$z$, perpendicular to the plane and passing through the intersection of $x$-$x$ and $y$-$y$ is given by the relation:

$$I_{zz} = I_{xx} + I_{yy} \tag{9.5}$$

**Proof**   A small lamina $L$ of area $\delta A$ having its coordinates $(x, y)$ along two mutually perpendicular axes $Ox$ and $Oy$ is considered in Figure 9.6. The plane $Oz$ is perpendicular to $Ox$ and $Oy$. Let $r$ be the distance of the small lamina $L$ from $zz$ axis, so that $OL = r$ and from geometry, $r^2 = x^2 + y^2$.

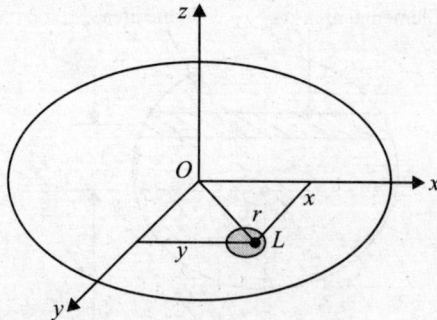

**Figure 9.6**   Proof of theorem of perpendicular axis.

Now moment of inertia of lamina $L$ about $x$-$x$ is:

$$I_{xx} = \delta A \cdot y^2$$

Similarly

$$I_{yy} = \delta A \cdot x^2$$

and

$$I_{zz} = \delta A \cdot r^2 = \delta A (x^2 + y^2)$$
$$I_{zz} = \delta A \cdot x^2 + \delta A \cdot y^2 = I_{yy} + I_{xx}$$

## 9.7    MOMENT OF INERTIA BY METHOD OF INTEGRATION

In Figure 9.7, an elemental area $\delta A$ of the plane area $A$ shown by shaded lines is considered.

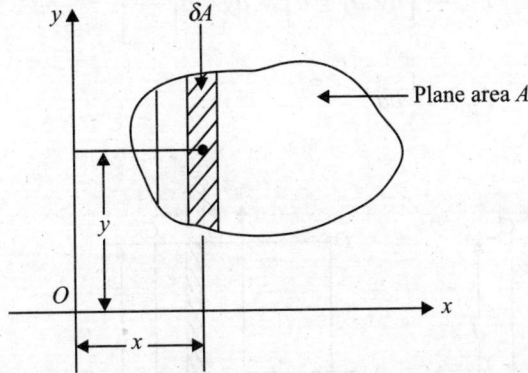

**Figure 9.7**    Moment of inertia by integration method.

Let $x$ be the distance of CG of this small area from $x$-axis.

Let $y$ be the distance of CG of this small area from $y$-axis.

Moment of inertia of this small area about $y$-axis = $\delta A \cdot x^2$

Moment of the whole area about $y$-axis = $I_{xx} = \sum \delta A \cdot x^2$      (9.6)

Similarly

$$I_{yy} = \sum \delta A \cdot y^2 \tag{9.7}$$

### 9.7.1    When the Section is Rectangular

In the rectangular section $ABCD$ shown in Figure 9.8(a), consider a small strip of thickness $\delta y$ which is at a distance $y$ from $x$-$x$.

Let the moment of inertia of the given section $ABCD$ about $x$-$x$ be $I_{xx}$. The moment of inertia of the elementary strip of area $b \, \delta y = (b \, \delta y) \cdot y^2$.

Now

$$I_{xx} = \int_{-d/2}^{d/2} by^2 dy = b \int_{-d/2}^{d/2} y^2 dy = b\left[\frac{y^3}{3}\right]_{-d/2}^{d/2} = \frac{b}{3}\left[\left(\frac{d}{2}\right)^3 - \left(-\frac{d}{2}\right)^3\right] \tag{9.8a}$$

or      $I_{xx} = \dfrac{1}{12}bd^3$      (9.8b)

Similarly, considering Figure 9.8(b) and proceeding in the same way,

$$I_{yy} = \frac{1}{12}db^3 \tag{9.9}$$

When the moment of inertia is taken about the base line $AB$ in Figure 9.8(c), the limit of integration will be from 0 to $d$ and moment of inertia can be worked out to be:

$$I_{AB} = \int_0^d by^2 dy = b\int_0^d y^2 dy = b\left[\frac{y^3}{3}\right]_0^d = \frac{1}{3}bd^3$$

or

$$I_{AB} = \frac{1}{3}bd^3 \tag{9.10}$$

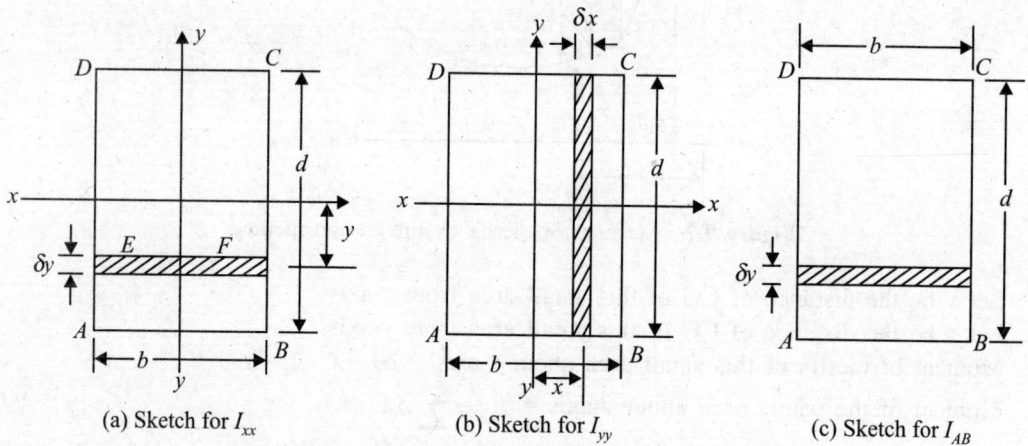

(a) Sketch for $I_{xx}$    (b) Sketch for $I_{yy}$    (c) Sketch for $I_{AB}$

**Figure 9.8**    Moment of inertia of a rectangular section.

## 9.7.2    Moment of Inertia of a Hollow Rectangular Section

Figure 9.9 shows a hollow rectangular section in which section $EFGH$ is cut out from the main section $ABCD$.

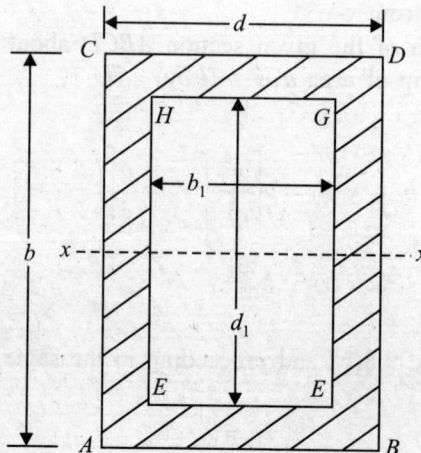

**Figure 9.9**    Moment of inertia of a hollow rectangular section.

Now moment of inertia of section $ABCD$ about $xx$ is $= \dfrac{1}{12}bd^3$

Moment of inertia of section $EFGH$ about $xx$ is $= \dfrac{1}{12}b_1d_1^3$

Hence, moment of inertia of the hollow section $ABCD$ about $xx$ is:

$$I_{xx} = \frac{1}{12}bd^3 - \frac{1}{12}b_1d_1^3 \qquad (9.11a)$$

Similarly

$$I_{yy} = \frac{1}{12}db^3 - \frac{1}{12}d_1b_1^3 \qquad (9.11b)$$

## 9.7.3   Moment of Inertia of a Circular Section

Figure 9.10 shows the circle $ABCD$ with radius $R$ or diameter $D$ and centre at $O$ with axes $x$-$x$ and $y$-$y$. Now consider an elemental ring of radius $r$ with thickness $dr$ as shown in the figure. The area of this circular ring is $da = 2\pi \cdot dr$ and moment of inertia of this ring about $x$-$x$ and $y$-$y$ axis

$$= \text{Area} \times (\text{distance})^2$$
$$= 2\pi r \cdot dr \times r^2$$
$$= 2\pi r^3 \cdot dr$$

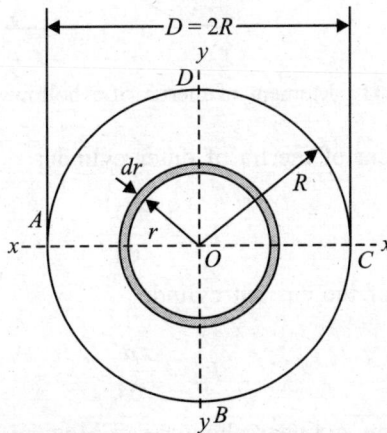

**Figure 9.10**   Moment of inertia of a circular section.

Moment of inertia of the whole circular section about an axis passing through $O$ and perpendicular to the pane of the paper (or $x$-$x$ or $y$-$y$)

$$= I_{zz} = \int_0^R 2\pi r^3 dr = 2\pi \int_0^R r^3 dr = 2\pi \left[\frac{r^4}{4}\right]_0^R = 2\pi \frac{R^4}{4} = \frac{\pi \left(\dfrac{D}{2}\right)^4}{2}$$

or
$$I_{zz} = \frac{\pi D^4}{32}$$
(9.12)

From the theorem of perpendicular axis, i.e. Eq. (9.5), $I_{zz} = I_{xx} + I_{yy}$, but due to symmetry

$$I_{xx} = I_{yy}$$

$$\therefore \qquad I_{xx} = I_{yy} \frac{I_{zz}}{2} = \frac{\pi D^4}{64}$$
(9.13)

### 9.7.4  Moment of Inertia of a Hollow Circular Section

Sectional view of a hollow circular cylinder is shown in Figure 9.11 with its external and internal diameters as $D$ and $d$ respectively.

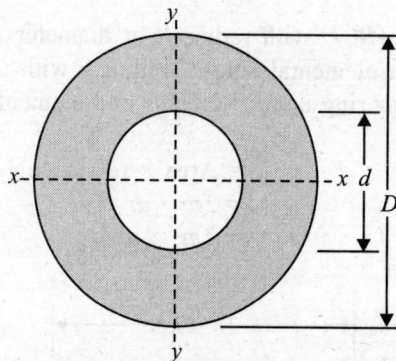

**Figure 9.11**   Moment of inertia of a hollow cylinder.

Applying Eq. (9.13), moment of inertia of outer cylinder

$$I_{xx} = \frac{\pi D^4}{64}$$

Similarly, moment of inertia of the cut out cylinder

$$I_{xx} = \frac{\pi d^4}{64}$$

Moment of inertia of the hollow cylinder about $x$-$x$ = Moment of inertia of outer cylinder – moment of inertia of the cut out cylinder, i.e.

$$I_{xx} = \frac{\pi D^4}{64} - \frac{\pi d^4}{64} = \frac{\pi}{64}(D^4 - d^4)$$
(9.14a)

Similarly

$$I_{yy} = \frac{\pi D^4}{64} - \frac{\pi d^4}{64} = \frac{\pi}{64}(D^4 - d^4) = I_{xx}$$
(9.14b)

## 9.7.5   Moment of Inertia of a Triangular Section About the Base

A triangular section $ABC$ whose base and height $b$ and $h$ are respectively considered in Figure 9.12. Consider a strip $EF$ of thickness $dy$ from a depth $y$ from vertex $A$. Now the triangles $AEF$ and $ABC$ are similar. Hence $\dfrac{EF}{BC} = \dfrac{y}{h}$

$$\therefore \qquad EF = BC\frac{y}{h} = b\frac{y}{h}$$

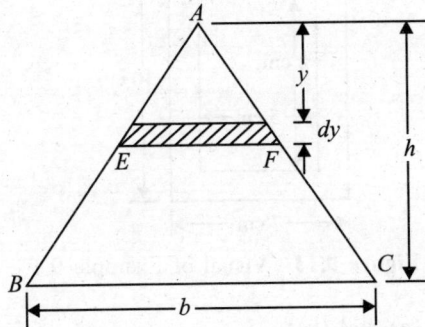

**Figure 9.12**   Moment of inertia of a triangular section.

Area of this strip $EF = b\dfrac{y}{h} \cdot dy$

Moment of inertia of this area of the strip about base $b$ = area $\times$ (distance)$^2$ = $b\dfrac{y}{h} \cdot (h-y)^2\, dy$

Moment of inertia of the whole section about the base $b$ is:

$$I_{BC} = \int_0^h b\frac{y}{h}(h=y)^2\, dy = \frac{b}{h}\int_0^h (h^2 y + y^3 - 2hy^2)dy = \frac{b}{h}\left[\frac{h^2 y^2}{2} + \frac{y^4}{4} - \frac{2hy^3}{3}\cdot\right]_0^h$$

or $\qquad I_{BC} = \dfrac{bh^3}{12}$ \hfill (9.15)

We know that the distance between the centre of gravity $G$ of the triangle and its $BC$ is $\dfrac{h}{3}$. Applying law of parallel axis theorem

$$I_{BC} = I_G + \text{Area of the triangle} \times \left(\frac{h}{3}\right)^2$$

$$I_{BC} = I_G + \frac{1}{2}bh \times \left(\frac{h}{3}\right)^2 = I_G + \frac{1}{18}bh^3$$

or

$$I_G = I_{BC} - \frac{1}{18}bh^3 = \frac{bh^3}{12} - \frac{bh^3}{18}$$

$\therefore$

$$I_G = \frac{bh^3}{36}$$ (9.16)

**EXAMPLE 9.1** The dimensions of a hollow rectangular section are given in Figure 9.13. Find the moment of inertia $I_{xx}$ and $I_{yy}$.

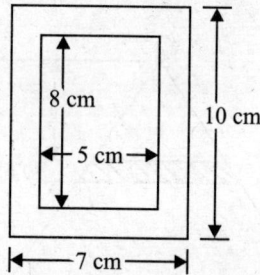

**Figure 9.13** Visual of Example 9.1.

**Solution** Use the Eqs. [9.11(a) and (b)]

$$I_{xx} = \frac{1}{12}bd^3 - \frac{1}{12}b_1d_1^3$$

$$I_{xx} = \frac{1}{12} \times 7 \times 10^3 - \frac{1}{12} \times 5 \times 8^3$$

$$I_{xx} = \frac{7000}{12} - \frac{2580}{12} = \frac{4420}{12} = 368.33 \, \text{cm}^4 \qquad \text{(Answer)}$$

$$I_{yy} = \frac{1}{12}db^3 - \frac{1}{12}d_1b_1^3$$

$$I_{yy} = \frac{1}{12} \times 10 \times 7^3 - \frac{1}{12} \times 8 \times 5^3$$

$$I_{yy} = \frac{3430}{12} - \frac{1000}{12} = \frac{2430}{12} = 202.5 \, \text{cm}^4 \qquad \text{(Answer)}$$

**EXAMPLE 9.2** The external and internal diameters of a hollow circular section are given in Figure 9.14. Find the moment of inertia about the horizontal axis passing through the CG.

**Solution** Use Eq. (9.14), i.e.

$$I_{xx} = I_{yy} = \frac{\pi D^4}{64} - \frac{\pi d^4}{64} = \frac{\pi}{64}(D^4 - d^4) = \frac{\pi}{64}(10^4 - 8^4)$$

$$I_{xx} = I_{yy} = \frac{\pi}{64}(10000 - 4096) = \frac{\pi \times 5904}{64} = 289.81 \, \text{cm}^4 \qquad \text{(Answer)}$$

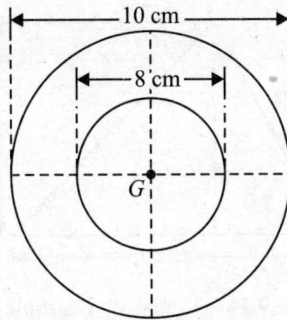

**Figure 9.14**  Visual of Example 9.2.

**EXAMPLE 9.3**  Find the moment inertia of the rectangular section shown in Figure 9.15 about the axis x-x through G and about BC.

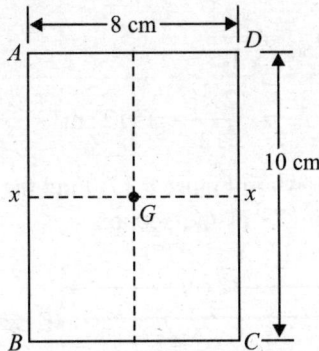

**Figure 9.15**  Visual of Example 9.3.

**Solution**  Using Eq. [9.8(b)]

$$I_{xx} = \frac{1}{12}bd^3$$

$$I_{xx} = \frac{1}{12} \times 8 \times 10^3 = 666.66 \text{ cm}^4 \qquad \text{(Answer)}$$

Using Eq. (9.10)

$$I_{BC} = \frac{1}{3}bd^3$$

$$I_{BC} = \frac{1}{3} \times 8 \times 10^3 = 2666.66 \text{ cm}^4 \qquad \text{(Answer)}$$

**EXAMPLE 9.4**  Determine the moment of inertia of the triangular section, shown in Figure 9.16, about centre of gravity and about the base BC.

**Solution**  Using Eq. (9.16),

$$I_G = \frac{bh^3}{36}$$

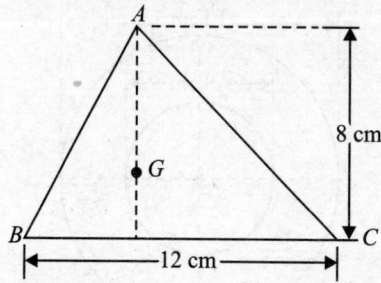

**Figure 9.16**   Visual of Example 9.4.

or

$$I_G = \frac{12 \times 8^3}{36} = 170.66 \text{ cm}^4 \qquad \text{(Answer)}$$

Using Eq. (9.15)

$$I_{BC} = \frac{bh^3}{12}$$

$$I_G = \frac{12 \times 8^3}{12} = 512 \text{ cm}^4 \qquad \text{(Answer)}$$

**EXAMPLE 9.5**   A T-section is shown in Figure 9.17. Find the moment of inertia of this section about *x-x* axis passing through the CG of the section.

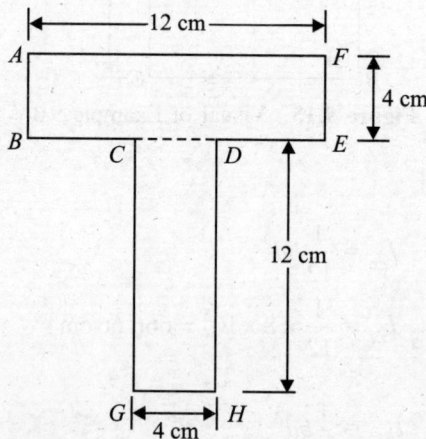

**Figure 9.17**   Visual of Example 9.5.

**Solution**   The T-section is symmetrical about *y-y* axis; therefore, CG of the section lies on this *y-y* axis. The T-section is split into two rectangles *ABCDEF* and *CDHG* as shown in Figure 9.17.

Let $\bar{y}$ be the distance between the CG and the top surface *AF* of the T-section.
Considering rectangle *ABCDEF*,

$$\text{Area} = A_1 = 12 \times 4 = 48 \text{ cm}^2 \qquad \text{and} \qquad y_1 = 12 + (4/2) = 14 \text{ cm}$$

Considering rectangle *CDHG*,

$$\text{Area} = A_2 = 12 \times 4 = 48 \text{ cm}^2 \quad \text{and} \quad y_2 = (12/2) = 6 \text{ cm}$$

Now using the equation

$$\bar{y} = \frac{A_1 y_1 + A_2 y_2}{A_1 + A_2} = \frac{48 \times 14 + 48 \times 6}{48 + 48} = 10 \text{ cm}$$

Distance between the CG of the given section and CG of the rectangle *ABCDEF*,

$$h_1 = y_1 - \bar{y} = 14 - 10 = 4 \text{ cm}$$

Distance between the CG of the given section and CG of the rectangle *CDHG*,

$$h_2 = \bar{y} - y_2 = 10 - 6 = 4 \text{ cm}$$

Now

$$I_{G1} = \frac{12 \times 4^3}{12} = 64 \text{ cm}^4$$

$$I_{G2} = \frac{4 \times 12^3}{12} = 576 \text{ cm}^4$$

Now applying theorem of parallel axis, moment of inertia of the rectangle *ABCDEF* about the horizontal axis passing through CG of the given section

$$= I_{G1} + A_1 h_1^2 = 64 + 48 \times 4^2 = 832 \text{ cm}^4$$

Similarly, moment of inertia of the rectangle *CDHG* about the horizontal axis passing through CG of the given section

$$= I_{G2} + A_2 h_2^2 = 576 + 48 \times 4^2 = 1344 \text{ cm}^4$$

∴ Moment of inertia of the given section about the horizontal axis passing through the CG of the given section

$$= I_{G1} + I_{G2} = 832 + 1344 = 2176 \text{ cm}^4 \quad \text{(Answer)}$$

Moment of inertia of the given section about the vertical axis passing through the CG of the given section

$$= \frac{4 \times 12^3}{12} + \frac{12 \times 4^3}{12} = 640 \text{ cm}^4 \quad \text{(Answer)}$$

**EXAMPLE 9.6** Find the moment of inertia of the L-section, shown in Figure 9.18, about centroidal *x-x* and *y-y* axes.

**Solution**

**Case I:** Moment of inertia about centroidal *x-x* axis.
Rectangle *ABCDE*

$$\text{Area} = A_1 = 12 \times 2 = 24 \text{ cm}^2, \quad y_1 = (12/2) = 6 \text{ cm}$$

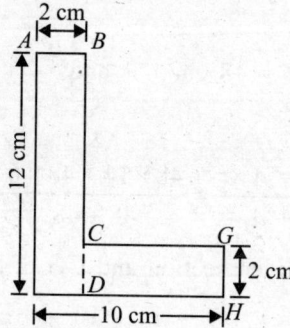

**Figure 9.18** Visual of Example 9.6.

Rectangle *CDHG*

$$\text{Area} = A_2 = (10 - 2) \times 2 = 16 \text{ cm}^2, \qquad y_2 = (2/2) = 1 \text{ cm}$$

Let $\bar{y}$ be the distance between the CG of the section and the bottom face.
Using the relation

$$\bar{y} = \frac{A_1 y_1 + A_2 y_2}{A_1 + A_2} = \frac{24 \times 6 + 16 \times 1}{24 + 16} = 4 \text{ cm}$$

Moment of inertia of rectangle *ABCDE* about an axis through CG and parallel to *x-x* axis is:

$$I_{G1} = \frac{2 \times 12^3}{12} = 288 \text{ cm}^4$$

Distance of CG of the rectangle *ABCDE* from *x-x* axis = $h_1 = y_1 - \bar{y} = 6 - 4 = 2$ cm
Moment of inertia of rectangle *ABCDE* about *x-x* axis = $I_{G1} + A_1 h_1^2$

$$= 288 + 24 \times 2^2 = 384 \text{ cm}^4$$

Similarly moment of inertia of rectangle *CDHG* about an axis through its CG and parallel to *x-x* axis:

$$I_{G2} = \frac{(10 - 2) \times 2^3}{12} = \frac{8 \times 8}{12} = 5.333 \text{ cm}^4$$

Distance of CG of rectangle *CDHG* from *x-x* axis = $h_2 = \bar{y} - y_2 = 4 - 1 = 3$ cm
Moment of inertia of the rectangle *CDHG* about *x-x* axis

$$= I_{G2} + A_2 h_2^2 = 5.333 + 16 \times 3^2 = 149.333 \text{ cm}^4$$

Now moment of inertia of the whole section about *x-x* axis is:

$$I_{xx} = 384 + 149.33 = 533.333 \text{ cm}^4 \qquad \text{(Answer)}$$

**Case II:** Moment of inertia about centroidal axis *y-y* axis.
Rectangle *ABCDE*

$$\text{Area} = A_1 = 12 \times 2 = 24 \text{ cm}^2, \qquad x_1 = (2/2) = 1 \text{ cm}$$

Rectangle *CDHG*

$$\text{Area} = A_2 = (10 - 2) \times 2 = 16 \text{ cm}^2, \qquad x_2 = 2 + (8/2) = 6 \text{ cm}$$

Now using the equation

$$\bar{x} = \frac{A_1 x_1 + A_2 x_2}{A_1 + A_2} = \frac{24 \times 1 + 16 \times 6}{24 + 16} = 2.8 \text{ cm}$$

Moment of inertia of rectangle *ABCDE* about an axis through CG and parallel to *y-y* axis is:

$$I_{G1} = \frac{12 \times 2^3}{12} = 8 \text{ cm}^4$$

Distance of CG of the rectangle *ABCDE* from *y-y* axis = $h_1 = 2.8 - 1 = 1.8$ cm

Moment of inertia of rectangle *ABCDE* about *y-y* axis = $I_{G1} + A_1 h_1^2$

$$= 8 + 24 \times 1.82 = 85.76 \text{ cm}^4$$

Similarly moment of inertia of rectangle *CDHG* about an axis through its CG and parallel to *y-y* axis:

$$I_{G2} = \frac{2 \times 8^3}{12} \, 85.333 \text{ cm}^4$$

Distance of CG of rectangle *CDHG* from *y-y* axis = $h_2 = 6 - 2.8 = 3.2$ cm

Moment of inertia of the rectangle *CDHG* about *y-y* axis

$$= I_{G2} + A_2 h_2^2 = 85.333 + 16 \times 3.2^2 = 249.173 \text{ cm}^4$$

Now moment of inertia of the whole section about *y-y* axis is:

$$I_{yy} = 85.76 + 249.173 = 334.933 \text{ cm}^4 \qquad \text{(Answer)}$$

**EXAMPLE 9.7** A hollow section with a circular hole is shown in Figure 9.19. Find the moment of inertia of the hollow section about an axis passing through its CG or parallel to *x-x* axis.

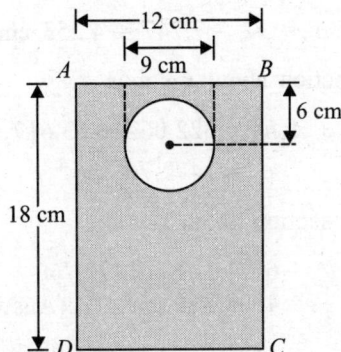

**Figure 9.19** Visual of Example 9.7.

*Solution*    It is seen in Figure 9.19 that the section is symmetrical about $y$-$y$ axis and, therefore, the CG of the section will lie on it. Let $\bar{y}$ be the distance between the CG of the section from the bottom face *CD*.

Considering the rectangle *ABCD*,

$$A_1 = 18 \times 12 = 216 \text{ cm}^2 \qquad y_1 = 18/2 = 9 \text{ cm}$$

Considering the circular whole

$$A_2 = \frac{\pi}{4} \times 9^2 = 63.617 \text{ cm}^2 \qquad y_2 = 18 - 6 = 12 \text{ cm}$$

Using with usual notations

$$\bar{y} = \frac{A_1 x_1 - A_2 x_2}{A_1 - A_2} = \frac{216 \times 9 - 63.617 \times 12}{216 - 63.617} = 7.747 \text{ cm}$$

Moment of inertia of the rectangular section about its axis through the CG and parallel to $x$-$x$ axis,

$$I_{G1} = \frac{12 \times 18^3}{12} = 5832 \text{ cm}^4$$

Distance of CG of rectangular section and $x$-$x$ axis

$$= h_1 = 9 - 7.747 = 1.253 \text{ cm}$$

Moment of inertia of rectangular section about $x$-$x$ axis

$$= I_{G1} + A_1 h_1^2 = 5832 + 216 \times 1.253^2$$
$$= 6171.122 \text{ cm}^4$$

Similarly, moment of inertia of the circular section about an axis through the CG and parallel to $x$-$x$ axis

$$I_{G2} = \frac{\pi}{64} \times 9^4 = 322.062 \text{ cm}^4$$

Distance of CG of circular section and $x$-$x$ axis

$$= h^2 = 12 - 7.747 = 4.253 \text{ cm}$$

Moment of inertia of circular section about $x$-$x$ axis

$$= I_{G2} + A_2 h_2^2 = 322.062 + 63.617 \times 4.253^2$$
$$= 1472.766 \text{ cm}^4$$

Moment of inertia of the whole section about $x$-$x$ axis

$$= I_{xx} = 6171.122 - 1472.766$$
$$= 4698.356 \text{ cm}^4 \qquad \text{(Answer)}$$

## 9.8    CONCLUSION

Moments of a force and plane area and then second moment or moment of inertia are presented with figures and equation at the beginning. Then radius of gyration, section modulus, theorem of parallel axis, theorem of perpendicular axis are defined and equations are derived. Equations of moment of inertia of different sections like rectangular, circular, triangular by method of integration are derived. Moment of inertia of different sections is essential in day-to-day use in engineering mechanics, strength of materials, structural engineering, etc. A few numerical problems are solved to clear the theoretical equations derived.

## EXERCISES

**9.1** Define moment of inertia and find out the moment of inertia of a plane area.

**9.2** What is radius of gyration and section modulus? Give their equations.

**9.3** State and proof theorem of parallel axis.

**9.4** State and proof theorem of perpendicular axis.

**9.5** Find the expressions of moment of inertia of rectangular and triangular section about their CG.

**9.6** Derive the equations of moment of inertia of hollow and rectangular sections about their CG.

**9.7** Find the moment of inertia of a rectangular section of 12 cm long and 10 cm deep about the CG.

(Answer    $I_{xx} = 1000$ cm$^4$, $I_{yy} = 1440$ cm$^4$)

**9.8** The base of a triangular section is 14 cm. If its perpendicular distance of the base from the vertex is 12 cm, find its moment of inertia about the CG.

(Answer    $I_{xx} = 672$ cm$^4$)

# CHAPTER 10

# Friction

## 10.1 INTRODUCTION

The surfaces of body are not truly smooth. If it is carefully observed through microscope, it will be seen that the surface which we roughly assume to be smooth, is found to have roughness and irregularity. Therefore, when an object or a surface begins to move under the action of some forces on another surface on which it was initially on rest, the roughness or the interlocking properties of the projecting particles offer some resistance opposing the motion. The force of resistance or opposing force on the moving body is called *frictional force* or simply *friction*. This friction acts parallel to the surface of contact and depends on the nature or intensities of roughness. Thus it may be concluded that frictional force is developed by virtue of resistance between the contact surfaces of two bodies when one moves or tends to move over the other. The friction is zero for pure smooth surfaces.

## 10.2 TYPES OF FRICTION

Various types of friction are:

   (i) Static friction
  (ii) Solid or dry friction
 (iii) Kinetic or dynamic friction
 (iv) Limiting friction
  (v) Fluid friction

Kinetic or dynamic friction is again divided into:

  (a) Sliding friction
  (b) Rolling friction

### 10.2.1 Static Friction

It is the friction experienced by a body when it is in rest. In other words, if the two surfaces which are in contact, are at rest, the force experienced by the one surface is called *static friction*.

### 10.2.2 Solid or Dry Friction

If between the two surfaces, no lubrication like oil or grease is used, the friction that exists between the two surfaces is called *solid* or *dry friction*.

### 10.2.3 Kinetic or Dynamic Friction

If one surface starts moving and other is at rest, the force experienced by the moving surface is called *dynamic* or *kinetic friction*. When a body slides over another surface or body, friction experienced by the body is called *sliding friction*. When balls or rollers are interposed between two surfaces, the friction experienced by the balls or rollers is called *rolling friction*.

### 10.2.4 Limiting Friction

If a body resting on another body or surface is gently pulled by force $P$ as shown in Figure 10.1, it does not move due to force of friction $F$ which prevents the motion. If the pull $P$ is slightly increased, body may be found to be in equilibrium. It shows that force friction $F$ has increased itself so as to become equal and opposite to applied force $P$. There is, however, a limit of force of friction $F$ and the force of pull $P$ just exceeds this limit, force of friction cannot balance and body just begins to move. The body just begins to move or slide at some maximum value of $F$. This maximum value of $F$ is known as limiting friction.

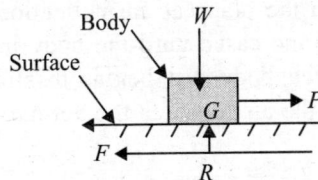

**Figure 10.1** Definition of limiting friction.

### 10.2.5 Fluid Friction

The friction between two fluid layers or the friction between a solid and a fluid is called *fluid friction*.

### 10.3 LAWS OF FRICTION

(i) The direction of force of friction is always opposite to that in which the point of contact has tendency to slide.

(ii) The magnitude of force of friction is such as would be just sufficient to prevent the sliding motion of the point of contact, subject to a certain maximum limit.

(iii) The magnitude of limiting friction always bears a constant ratio to the normal pressure between the two bodies in contact and this ratio depends only on the nature of roughness of the materials of which the bodies are composed of, but not on the shape or extent of the surfaces in contact.

(iv) The constant ratio between the magnitudes of kinetic friction to the normal reaction is slightly less than that in case of limiting friction.

(v) For moderate speed, the force of friction remains constant. It decreases slightly with the increase of speed.

## 10.4   ANGLE OF FRICTION

Consider Figure 10.2, where body is resting on an inclined plane.

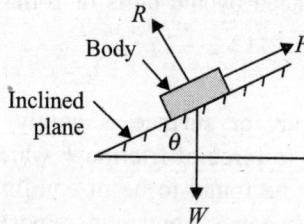

**Figure 10.2**   Sketch to define angle of friction $\phi$.

Let $W$ be the weight of the acting vertically downwards, normal reaction of the inclined plane is $R$ and $F$ is frictional force acting upwards parallel to the inclined plane.

The body is in equilibrium on the plane at an inclination of $\theta$ to the horizontal. Let this angle of inclination gradually be increased until the body just starts sliding down the plane. This angle of inclination at which body just begins to slide is called *angle of friction* $\phi$. This angle of friction is equal to the angle which the normal reaction makes with the vertical.

## 10.5   COEFFICIENT OF FRICTION

It is defined as the ratio of the limiting force of friction $F$ to the normal reaction $R$ between the two bodies and denoted by $\mu$. Thus:

$$\mu = \frac{\text{Limiting force of friction}}{\text{Normal reaction}} = \frac{F}{R} = \tan \phi \qquad (10.1)$$

$\therefore$
$$F = \mu R \qquad (10.2)$$

**EXAMPLE 10.1**   A pull of 180 N at an angle of 30° to the rough horizontal plane is required to just to move a body of weight $W$ as shown in Figure 10.3(a). It was found that a push of 200 N at an angle of 30° to the plane is just enough to remove the body as shown in Figure 10.3(b). Find the weight $W$ and coefficient of friction $\mu$.

**Solution**   From Figure 10.3(a), resolving the forces horizontally,

$$F = 180 \cos 30° = 180 \times 0.866 = 155.88 \text{ N}$$

Resolving the forces vertically,

$$R = W - 180 \sin 30° = (W - 90) \text{ N}$$

Using Eq. (10.2), $F = \mu R$
or
$$155.88 = \mu(W - 90) \qquad\qquad\qquad\qquad \text{(i)}$$

From Figure 10.3(b), resolving the forces horizontally,

$$F = 200 \cos 30° = 200 \times 0.866 = 173.2 \text{ N}$$

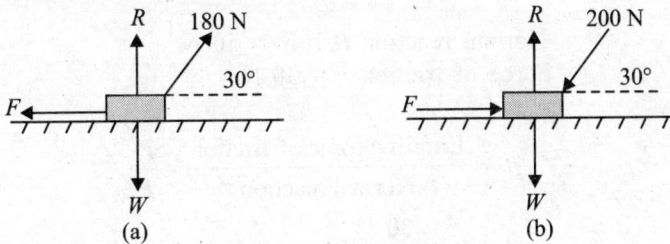

**Figure 10.3**   Visual of Example 10.1.

Resolving the forces vertically,

$$R = W + 200 \sin 30° = (W + 100) \text{ N}$$

Using Eq. (10.2), $F = \mu R$
or
$$173.2 = \mu(W + 100) \qquad\qquad\qquad\qquad \text{(ii)}$$

Dividing (i) by (ii),

$$\frac{155.88}{173.2} = \frac{\mu(W - 90)}{\mu(W + 100)} = \frac{W - 90}{W + 100}$$

$$0.9 = \frac{W - 90}{W + 100}$$

$$0.9 \; W + 90 = W - 90$$

or
$$W - 0.9 \; W = 90 + 90 = 180$$

$$\therefore \qquad\qquad W = 1800 \text{ N} \qquad \text{(Answer)}$$

Substituting the value of $W$ in Eq. (i),

$$155.88 = \mu(1800 - 90)$$

$$\therefore \qquad\qquad \mu = \frac{155.88}{1800 - 90} = \frac{155.88}{1710} = 0.09115 \qquad \text{(Answer)}$$

**EXAMPLE 10.2**   A horizontal force of 30 N just causes a body of weight 50 N to move on a horizontal rough plane. Determine the coefficient of friction $\mu$.

*Solution*   Consider Figure 10.4.

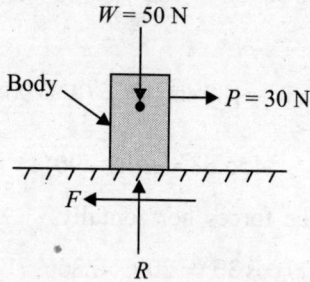

**Figure 10.4**   Visual of Example 10.2.

|  | Normal reaction $R = W = 50$ N |
|---|---|
| and | Force of friction $F = 30$ N |

Using Eq. (10.1)

$$\mu = \frac{\text{Limiting force of friction}}{\text{Normal reaction}} = \frac{F}{R}$$

$$\mu = \frac{F}{R} = \frac{30}{50} = 0.6 \qquad \text{(Answer)}$$

**EXAMPLE 10.3**   A body of weight 100 N is placed on rough horizontal plane. A force of 50 N is required at an angle of 20° to the horizontal just to cause the motion of the body. Find the coefficient of friction $\mu$.

*Solution*   As seen in Figure 10.5, force of friction $F = \mu R$ according to Eq. (10.2).

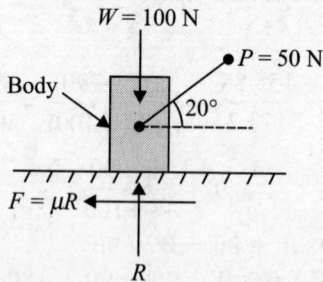

**Figure 10.5**   Visual of Example 10.3.

Resolving the forces along the plane or horizontally,

$$F = 50 \cos 20°$$

or
$$\mu R = 50 \cos 20° = 46.9846 \text{ N} \tag{i}$$

Resolving the forces normal to the plane or vertically,

$$R + 50 \sin 20° = 100 \text{ N}$$

or
$$R = 100 - 50 \sin 20° = 100 - 17.101 = 82.899 \text{ N}$$

Substituting the value of $R$ in (i),

$$\mu \times 82.899 = 46.9846$$

$\therefore$
$$\mu = \frac{46.9846}{82.899} = 0.5667 \qquad \text{(Answer)}$$

**EXAMPLE 10.4**  A body of weight $W$ is placed on a rough horizontal plane. A force $P$ is applied on the body at angle of $\theta$ with the horizontal as shown in Figure 10.6 such that the body just begins to slide. Prove that the applied force $P$ will be least, if angle $\theta$ is equal to the angle of friction $\phi$.

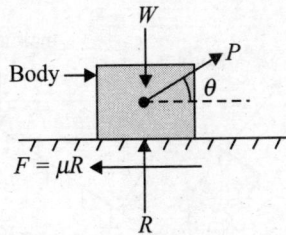

**Figure 10.6**  Visual of Example 10.4.

**Solution**  In Figure 10.6, the forces acting on the body are shown.
Resolving the forces normal to the plane or vertically,

$$R + P \sin \theta = W$$
$$R = W - P \sin \theta \qquad \text{(i)}$$

Resolving the forces along the plane or horizontally,

$$F = \mu R = P \cos \theta \qquad \text{(ii)}$$

Substituting the value of $R$ from (i) in (ii),

$$F = \mu R = P \cos \theta = \mu (W - P \cos \theta)$$

or
$$P \cos \theta = \mu (W - P \cos \theta)$$

Equation (10.1) gives

$$\mu = \tan \phi$$

Hence

$$P \cos \theta = \tan \phi \times (W - P \cos \theta)$$

or
$$P \cos \theta = \frac{\sin \phi}{\cos \phi} (W - P \cos \theta)$$

or
$$P \cos \theta \sin \phi = W \sin \phi - P \cos \theta \sin \phi$$

or
$$P \cos \theta \sin \phi + P \cos \theta \sin \phi = W \sin \phi$$

or
$$P \cos(\theta - \phi) = W \sin \phi$$

or
$$P = \frac{W \sin \phi}{\cos(\theta - \phi)} \qquad \text{(iii)}$$

The force $P$ will be least if the denominator of Eq. (iii) becomes, i.e. $\cos(\theta - \phi)$ is maximum. But $\cos(\theta - \phi)$ is maximum if $\cos(\theta - \phi) = 1$

or
$$(\theta - \phi) = 0$$
$$\therefore \qquad \theta = \phi \qquad \text{Proved.}$$

***EXAMPLE 10.5***  A solid body is placed on a plane inclined at an angle of $\alpha$ with horizontal as shown in Figure 10.7. Prove that this angle $\alpha$ is equal to angle of friction $\phi$ when the solid body starts moving downwards.

***Solution***  As seen in Figure 10.7, the body is on inclined plane $AC$ at an angle of $\alpha$ to the horizontal plane $AB$. The body is now under the action of the following forces.

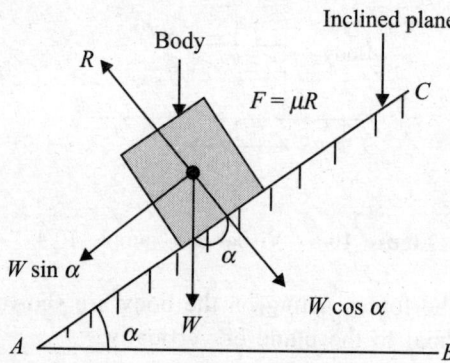

**Figure 10.7**  Visual of Example 10.5.

Weight of the body $W$ acting vertically downwards
Normal reaction $R$ acting perpendicular to the inclined plane $AC$
Frictional force $F = \mu R$ acting upwards along the inclined plane $AC$

Now the weight $W$ is resolved into two components $W \sin \alpha$ and $W \cos \alpha$ as shown in Figure 10.7.

Resolving the forces along and perpendicular to the inclined plane $AC$ respectively,

$$W \sin \alpha = F = \mu R \qquad \text{(i)}$$
and
$$W \sin \alpha = R \qquad \text{(ii)}$$

Dividing (i) by (ii) gives

$$\frac{W \sin \alpha}{W \cos \alpha} = \frac{\mu R}{R}$$

or
$$\tan \alpha = \mu$$

But Eq. (10.2) gives

$$\tan \phi = \mu$$
$$\therefore \qquad \tan \alpha = \tan \phi$$
$$\therefore \qquad \alpha = \phi \qquad \text{Proved.}$$

**EXAMPLE 10.6**   Two blocks $A$ and $B$ are connected by a horizontal rod and supported on two rough planes as shown in Figure 10.8. The coefficient of friction of the blocks $A$ and $B$ are 0.2 and 0.3. If the block $B$ weighs 150 N, what will be the weight of block $A$ for equilibrium of the system?

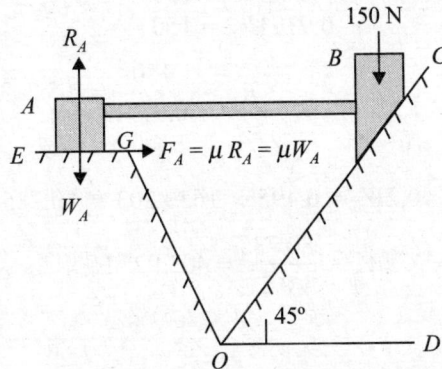

**Figure 10.8**   Visual of Example 10.6.

**Solution**   Considering Figure 10.8,

$W_B$ = 150 N, plane $OC$ is inclined 45° to the horizontal, $\mu_A$ = 0.2 and $\mu_B$ = 0.3, smallest weight of block $A$ is considered to be $W_A = R_A$. Frictional force along the surface $EG$ is equal to $F_A = \mu R_A = \mu W_A = 0.2\, W_A$ and this frictional force is transmitted to block $B$ horizontally as shown in Figure 10.8.

Figure 10.9 shows the forces acting on block $B$ and the components of frictional force $F_B$ and reaction $R_B$.

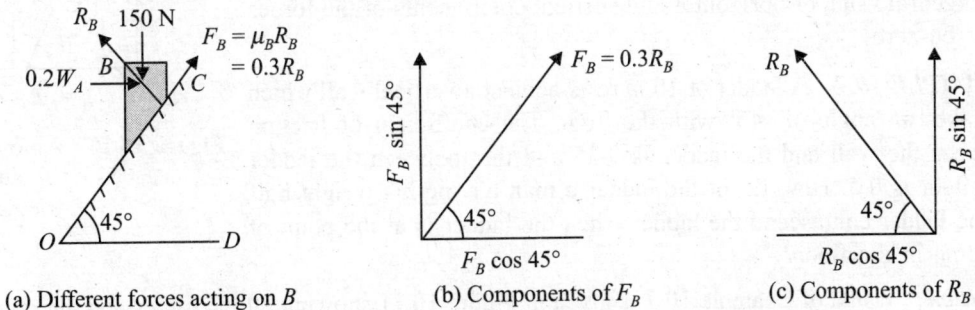

(a) Different forces acting on $B$     (b) Components of $F_B$     (c) Components of $R_B$

**Figure 10.9**   Visual of forces and components of Example 10.6.

Resolving the forces horizontally from Figure 10.9,

$$0.2W_A + F_B \cos 45° = R_B \cos 45°$$
$$0.2W_A + 0.3R_B \cos 45° = R_B \cos 45°$$
$$0.2W_A + 0.212R_B = 0.707R_B$$
$$0.2W_A = 0.707R_B - 0.212R_B$$
$$0.2W_A = 0.495R_B \tag{i}$$

Resolving the forces vertically,

$$F_B \sin 45° + R_B \sin 45° = 150$$
$$0.3R_B \sin 45° + R_B \sin 45° = 150$$
$$0.3R_B \times 0.707 + R_B \times 0.707 = 150$$
$$0.9191R_B = 150$$

$$\therefore \qquad R_B = \frac{150}{0.9191} = 163.203 \text{ N}$$

Substituting the value $R_B$ in (i),

$$0.2W_A = 0.495 \times 1630.203 = 80.785$$

$$\therefore \qquad W_A = \frac{80.785}{0.2} = 403,925 \text{ N} \qquad \text{(Answer)}$$

## 10.6   LADDER FRICTION

In Figure 10.10, a ladder $AB$ rests on the floor leaning against a vertical wall. The upper end of the ladder tends to slip downwards due to its self weight plus the weight of the person who climbs the ladder. Therefore, direction of the force of friction $F_B = \mu R_B$ between the wall and the ladder acts upwards as shown in Figure 10.10. Similarly, as the lower end of the ladder tends to slip away from the wall, direction of the force of friction $F_A = \mu R_A$ between the floor and ladder acts towards the vertical wall. As the system is in equilibrium, the algebraic sum of horizontal and vertical components of the forces must be zero.

**Figure 10.10**   Analysis of ladder friction.

**EXAMPLE 10.7**   A ladder of 10 m rests against a vertical wall which it makes an angle of 45° with the floor. The coefficient of friction between the wall and the ladder is 0.35 and that between the ladder and floor is 0.5. How far of the ladder a man having his weight half of the ladder can ascend the ladder when the ladder is at the point of slipping in the floor?

**Solution**   Visual of Example 10.7 is given in Figure 10.11 showing all the forces and directions, ladder of 10 m long, position and weight $W/2$ of the man at $M$ at a distant of $x$ from the floor, angle of inclination 45° of the ladder with the wall.

Now resolving the forces horizontally,

$$R_B = 0.5R_A$$
$$\therefore \qquad R_A = 2R_B \qquad \qquad \text{(i)}$$

Resolving the forces vertically,

$$R_A + 0.35R_B = W_L + \frac{W_L}{2} = \frac{3}{2}W_L \qquad \qquad \text{(ii)}$$

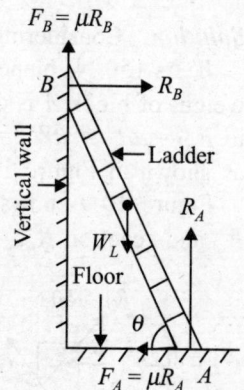

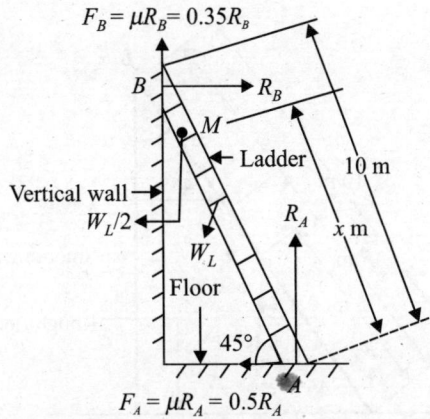

**Figure 10.11** Visual of Example 10.7.

Substituting the value of $R_A$ in (ii) from (i),

$$2R_B + 0.35R_B = 1.5W_L$$

or

$$(2 + 0.35)R_B = 1.5W_L$$

or

$$R_B = \frac{1.5}{2.35}W_L$$

$$R_B = 0.6383W_L \tag{iii}$$

Taking the moment of all the forces about $A$,

$$W_L \times 5\cos 45° + \frac{W_L}{2} \times x\cos 45° = 0.35R_B \times 10\cos 45° + R_B \times 10\sin 45°$$

$$3.5355W_L + 0.3535W_L x = 0.35 \times 0.6383W_L \times 10\cos 45° + 0.6383W_L \times 10\sin 45°3$$

or

$$3.5355 + 0.3535x = 0.35 \times 0.6383 \times 10\cos 45° + 0.6383 \times 10\sin 45°$$

or

$$0.3535x = 1.5547 + 4.5134 - 3.5355 = 2.5326$$

$$\therefore \qquad x = \frac{2.5326}{0.3535} = 7.1643 \text{ m} \qquad \text{(Answer)}$$

**EXAMPLE 10.8**   A ladder of 10 m long rests on a horizontal rough surface leaning against a smooth wall at an angle of 45° to the horizontal. The self-weight of the ladder is 800 N which acts through the mid-point of the ladder. A man weighing 650 N while climbing the ladder to the height of 7 m, the ladder is at the point of sliding. Determine the coefficient friction $\mu_f$ between the ladder and the rough floor.

**Solution**   In Figure 10.12, the ladder $AB$ of length 10 m is shown to lean against the smooth wall $CB$ resting on the rough floor $AC$ at an angle of 45° with horizontal. The self-weight of the ladder 800 N is acting vertically downwards through its mid-point and weight 650 N of the man also acts vertically downwards through a point 3 m below the top edge of the ladder. The reaction $R_A$ perpendicular to wall at $A$ and frictional force of the rough wall $F_f$ acts along the floor in the direction of $C$.

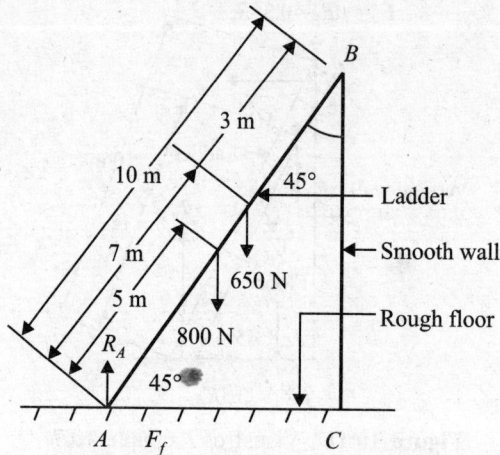

**Figure 10.12**   Visual of Example 10.8.

Let $\mu_f$ be the coefficient of friction between the rough floor and the ladder.
Now resolving the forces vertically,

$$R_A = (800 + 650) \text{ N} = 1450 \text{ N} \qquad \text{(i)}$$

Again

$$F_f = \mu_f R_A = \mu 1450 \text{ N} \qquad \text{(ii)}$$

Taking moment of the clockwise and anti-clockwise moments of the forces about $B$,

$$R_A \times 10 \sin 45° = F_f \times 10 \cos 45° + 650 \times 3 \sin 45° + 800 \times 5 \sin 45° \qquad \text{(iii)}$$

Substitute the values $R_A$ and $F_f$ from (i) and (ii) in (iii)

$$1450 \times 10 \sin 45° = \mu_f \times 1450 \times 7.071 + 650 \times 2.121 + 800 \times 3.535$$

or

$$10253.048 = \mu_f \times 10253.048 + 1378.65 + 2828$$

or  $\quad 10253.048 - 1378.65 - 2828 = 10253.048 \mu_f$

or  $\qquad\qquad 10253.048 \mu_f = 6046.398$

$\therefore \qquad\qquad\qquad \mu_f = \dfrac{6046.398}{10253.048} = 0.5897 \qquad$ **(Answer)**

## 10.7   WEDGE FRICTION

Figure 10.13 shows a wedge below a body. It is used either to lift a body or to adjust the position of the body for tightening fits or shafts. It is piece of metal or wood usually triangular in cross section as shown in Figure 10.13. As shown in Figure 10.13, a load $P$ is applied on the wedge. The angle $\theta$ is the angle of the wedge. It moves towards left, sliding occurs along the surfaces $BA$ and $CA$. The body moves upwards sliding along $EF$. Thus force of friction acts along the direction of $AB$, $AC$ and $EF$ as shown in Figure 10.13. Hence there will be three

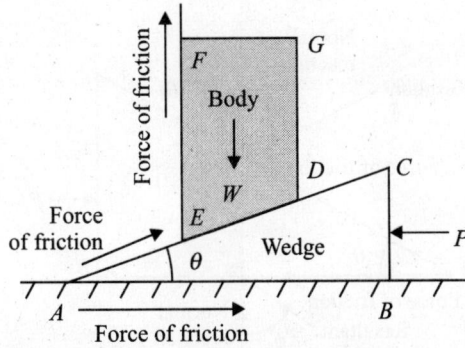

**Figure 10.13**  A wedge below a body to be lifted.

normal reactions at *AB*, *AC* and *EF*. Problems in wedge friction are solved by equilibrium methods or by Lami's theorem.

## 10.7.1  Analysis of Forces When the Body is in Equilibrium

Consider first the body *DEFG* to be in equilibrium in Figure 10.14

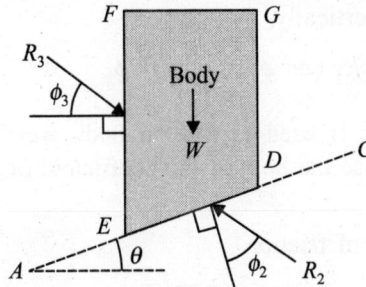

**Figure 10.14**  Forces $W$, $R_2$, $R_3$ acting on the body.

Here $W$ is the weight of the body acting vertically downwards, $R_2$ is the resultant of reaction force and force of friction on *AC*, $R_3$ is another resultant of reaction force and frictional force on *EF*. The angles $\phi_2$ and $\phi_3$ are the angles of friction along *AC* and *EF*.

Resolving the forces horizontally for equilibrium:

$$R_3 \cos \phi_3 = R_2 \sin (\phi_2 + \theta) \qquad (10.3)$$

Similarly resolving vertically,

$$R_3 \sin \phi_3 + W = R_2 \cos (\phi_2 + \theta) \qquad (10.4)$$

## 10.7.2  Analysis of Forces When the Wedge is in Equilibrium

Figure 10.15 shows different forces (normal and resultant on surface *AB* and *AC*) on the wedge which is in equilibrium.

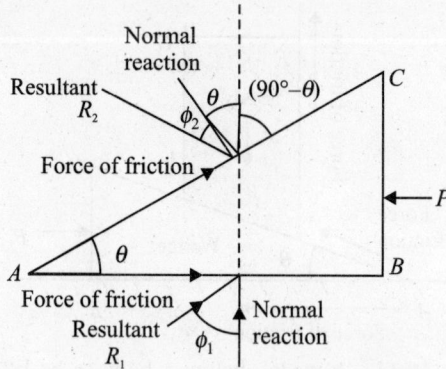

**Figure 10.15**    Forces on wedge in equilibrium.

Let $\phi_1$ and $\phi_2$ be the angles of friction at surface $AB$ and $AC$ and $\theta$ is the wedge angle as shown with resultants $R_1$ and $R_2$.

Now resolving the forces horizontally for equilibrium,

$$R_1 \sin \phi_1 + R_2 \sin(\phi_2 + \theta) = P \tag{10.5}$$

Similarly resolving the forces vertically,

$$R_1 \cos \phi_1 = R_2 \cos(\phi_2 + \theta) \tag{10.6}$$

**EXAMPLE 10.9**    A 10° wedge is used to raise a body weighing 1000 N. Determine the minimum force $P$ required to raise the body if the coefficient of friction $\mu = 0.35$ between all the surfaces.

**Solution**    The given coefficient of friction

$$\mu = 0.35$$

Hence

$$\tan \phi = 0.35$$
$$\phi = 19.29$$

Again the given wedge angle

$$\theta = 10°$$

Weight of the body

$$W = 1000 \text{ N}$$

Considering the equilibrium of the body (Figure 10.14), Eq. (10.3) is:

$$R_3 \cos \phi = R_2 \sin(\phi + \theta). \text{ Since } \phi_3 = \phi_2 = \phi_1 = \phi$$

or          $R_3 \cos 19.29° = R_2 \times \sin(19.29° + 10°) = R_2 \sin 29.29$

or                      $R_3 \times 0.9438 = R_2 \times 0.4992$

or                $R_2 = \dfrac{0.9438}{0.4838} R_3 = 1.9508 R_3$ 　　　　　　　　(i)

Using Eq. (10.4),

$$R_3 \sin \phi + W = R_2 \cos(\phi + \theta)$$

or $\qquad R_3 \sin 19.29° + 1000 = R_2 \cos(19.29° + 10°)$

or $\qquad R_3 \times 0.3303 + 1000 = R_2 \cos(29.29) = R_2 \times 0.8721$
$$= 1.9508 \, R_3 \times 0.8721 = 1.703 \, R_3$$

or $\qquad R_3(1.703 + 0.3303) = 1000$

or $\qquad R_3 = \dfrac{1000}{2.0333} = 491.8113 \text{ N}$

$\therefore \qquad R_2 = 1.9508 R_3 = 1.9508 \times 491.8113 = 959.425 \text{ N}$ $\qquad$ (ii)

Now consider the equilibrium of the wedge.
Using Eq. (10.6) from Figure 10.15,

$$R_1 \cos \phi + W = R_2 \cos(\phi + \theta)$$

or $\qquad R_1 \cos 19.29° + 1000 = R_2 \cos(19.29° + 10°)$
$$= 959.425 \times \cos 29.29°$$

or $\qquad R_1 \times 0.943 = 959.425 \times 0.8721 = 836.7145$

$$R_1 = \dfrac{836.7145}{0.9438} = 886.5379 \text{ N} \qquad \text{(iii)}$$

Using Eq. (10.5),

$$R_1 \sin \phi + R_2(\sin\phi + \theta) = P$$

or $\quad 996.5379 \times \sin 19.29° + 959.425 \times \sin(19.29° + 10°) = P$
$$P = 292.8234 + 469.3507$$
$$= 762.1741 \text{ N} \qquad \text{(Answer)}$$

## 10.8 SCREW JACK

The screws, bolts, nuts, etc. are widely used in machines and structures for fastenings. They have screw threads which are made by cutting a continuous helical groove on a cylindrical surface. Some of the common terms related to study screws or screw jack are:

**Helix:** It is the curve traced by particle while moving along a screw thread.

**Pitch:** It is the distance from a point of a screw to a corresponding point on the next thread measured parallel to the axis of the thread.

**Lead:** It is the distance a screw thread advances axially in one turn.

**Slope of the thread:** It is the inclination of thread with horizontal angle. If $\alpha$ is the inclination of the thread with horizontal or helix angle, then

$$\tan \alpha = \frac{\text{Lead of screw}}{\text{Circumference of screw}} = \frac{\text{Pitch}}{\pi d}$$

The screw jack is a device for lifting the heavy loads by applying comparatively smaller effort at its handle. The principle of screw jack is similar to an inclined plane. The load to be lifted is placed on the screw jack and the load is lifted by the application of effort at the end of the lever. Since the principle on which the jack works is similar to that of an inclined, the force applied on the lever of the screw jack is considered horizontal as shown in Figure 15.16.

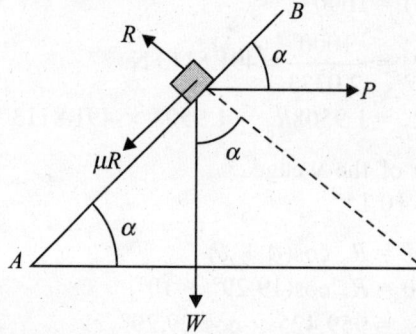

**Figure 10.16**    Forces when screw jack is in action.

Now resolving the forces along the plane $AB$

$$W \sin \alpha + \mu R = P \cos \alpha$$

or    $$W \sin \alpha + \tan \phi R = P \cos \alpha \qquad \text{(i)} \qquad \because \mu = \tan \phi$$

Resolving the forces perpendicular to the plane $AB$

$$R = W \cos \alpha + P \sin \alpha$$

Substituting the value of $R$ in (i)

$$P \cos \alpha = W \sin \alpha + \tan \phi (W \cos \alpha + P \sin \alpha)$$
$$P \cos \alpha = W \sin \alpha + W \tan \phi \cos \alpha + P \tan \phi \sin \alpha$$

or    $$P \cos \alpha - P \tan \phi \sin \alpha = W \sin \alpha + W \tan \phi \cos \alpha$$

or    $$P (\cos \alpha - \tan \phi \sin \alpha) = W (\sin \alpha + \tan \phi \cos \alpha)$$

$$\therefore \qquad P = W \times \frac{(\sin \alpha + \tan \phi \cos \alpha)}{(\cos \alpha - \tan \phi \sin \alpha)}$$

Now multiply the numerator and denominator by $\cos \phi$,

$$P = W \times \frac{(\sin \alpha \cos \phi + \tan \phi \cos \phi \cos \alpha)}{(\cos \alpha \cos \phi - \tan \phi \cos \phi \sin \alpha)}$$

or    $$P = W \times \frac{(\sin \alpha \cos \phi + \sin \phi \cos \alpha)}{(\cos \alpha \cos \phi - \sin \phi \sin \alpha)} = W \times \frac{\sin(\alpha + \phi)}{\cos(\alpha + \phi)}$$

$$\therefore \qquad P = W \tan(\alpha + \phi) \qquad \qquad \text{(10.7)}$$

Similarly it can be shown the relation between the effort $P$ and the weight $W$ lowered by a screw jack as:

$$P = W \tan(\phi - \alpha) \qquad \qquad \text{(10.8)}$$

**EXAMPLE 10.10**   A screw press s used to compress books. The thread is a double thread square head with a pitch of 5 mm and mean radius of 30 mm. The coefficient of friction $\mu$ for the contact surfaces of the threads is 0.35. Determine the torque required for a pressure of 650 N.

**Solution**   Given values are:
$p$ = 5 mm, radius $r$ = 30 mm, $W$ = 650 N, $m$ = 0.35, hence

$$\tan\phi = 0.35 \quad \text{or} \quad \phi = 19.29°$$

Let $P$ be the effort required at the mean radius of the screw for the torque and $\alpha$ be the helix angle.

$$\tan\alpha = \frac{2p}{\pi d} = \frac{2p}{2\pi r} = \frac{p}{\pi r} \quad \because \text{ double thread}$$

or

$$\tan\alpha = \frac{5}{\pi \times 30} = 3.0367°$$

Using Eq. (10.7),

$$P = W\tan(\alpha + \phi) = 650 \times \tan(19.29° + 3.0367°)$$

or      $P = 650 \times \tan(19.29° + 3.0367°) = 650 \times \tan 22.3267° = 266.938$ N

Hence torque $T$ required at mid-radius is:

$$T = P \times r = 266.938 \times 30 = 8008.14 \text{ N mm} \quad \text{(Answer)}$$

## 10.9   CONCLUSION

It is concluded from the above discussion that laws of friction, their applications and uses in different situations of engineering mechanics are important. In order to estimate the force to be applied in the field of climbing in a ladder, different lifting machines like applications of wedge, screw jack, pulleys, etc., determination of friction coefficient, angle of friction are necessary.

## EXERCISES

**10.1** What are the different types of friction? Give a brief description of each of them.

**10.2** State laws of friction, angle of friction and coefficient of friction.

**10.3** Explain with figures the ladder and wedge friction.

**10.4** A pull of 18 N is applied to a body weighing 50 N at an angle of 14° horizontally along a rough surface. Find the coefficient of friction.                    (Answer   0.382)

**10.5** A ladder of 7 m long rests on a rough horizontal floor and leans against a smooth vertical wall at an angle of 60° horizontally. The self-weight of the ladder is 800 N which acts through the middle of it. When a man weighing 600 N stands on the ladder at a distance of 4 m from the top of the ladder, the ladder is about to slide. Find the coefficient of friction between the ladder and the floor.          (Answer   0.237)

**10.6** Block *A* weighing 20 N is a rectangular prism resting on a rough inclined plane as shown in Figure 10.17. The block is tied by a horizontal string which has a tension of 6 N. Find
  (i) the frictional force on the block
  (ii) normal reaction of the inclined plane
  (iii) the coefficient of friction between the surfaces of contact.

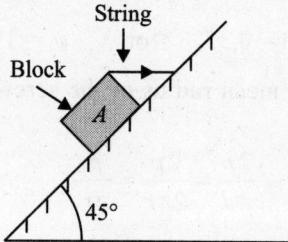

**Figure 10.17**    Visual of Exercise 10.6.

[Answer   (i) 9.898 N, (ii) 18.382 N, (iii) 0.538]

**10.7** A block weighing 1000 N is to be raised by means of 20° wedge as shown in Figure 10.18. Find the horizontal force *P* which will just raise the block if the angles of friction at the contact surfaces *AC*, *AB* and *DE* are 11°, 14° and 10°, respectively.

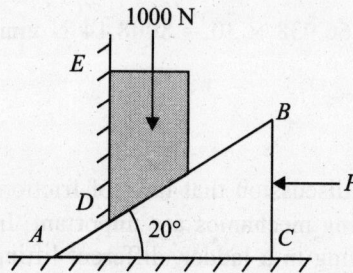

**Figure 10.18**    Visual of Exercise 10.7.

(Answer   1077 N)

# CHAPTER 11

# Kinetics of Rigid Bodies

## 11.1 INTRODUCTION

*Kinetics* is another branch of engineering in which the forces causing the motion, are not considered. This force produces acceleration on the rigid body causing its motion in the direction of force. Each and every body obeys some laws whether it is at rest or motion on straight path or rotating about an axis. These laws are called *Newton's Laws of Motion.*

## 11.2 NEWTON'S THREE LAWS OF MOTION

The first law states that a body continues in its state of rest or of uniform motion in a straight line, unless it is acted upon by an external force.

Newton's second law of motion states that the rate of change of momentum is directly proportional to the impressed force and takes place in the same direction in which the force acts.

The third law states that for every action, there is always an equal and opposite reaction.

## 11.3 SOME COMMON TERMS AND EQUATIONS

**Mass:**  The matter contained a body is called *mass.*

**Force:**  It is an agent which produces or tends to produce, destroy or tends to destroy motion.

**Velocity:**  It is the rate of change of moving body. If $s$ is the distance moved by a body in time $t$, the velocity $v$ is defined as:

$$v = \frac{ds}{dt} \qquad (11.1)$$

**Acceleration:**  It is the rate of change velocity. If $a$ is the acceleration then acceleration is written as:

$$a = \frac{dv}{dt} = \frac{d}{dt}\left(\frac{ds}{dt}\right) = \frac{d^2 s}{dt^2} \qquad (11.2)$$

This $\frac{dv}{dt}$ can also be written as:

$$\frac{dv}{dt} = \frac{dv}{ds} \cdot \frac{ds}{dt} = \frac{dv}{ds} \cdot v = v\frac{dv}{ds} \qquad (11.3)$$

**Momentum:**   It is the total motion possessed by a body. Mathematically if $M$ is the momentum, $m$ and $v$ is the velocity of the body, momentum is:

$$M = mv \qquad (11.4)$$

**Inertia:**   Inertia is an inherent property of a body which offers reluctance to change its state of rest or uniform motion.

**Work:**   It is the product of force $F$ and distance $s$ in the direction of force. Thus work done (WD), is expressed as:

$$WD = \text{Force} \times \text{distance} = F \cdot s \qquad (11.5)$$

**Power:**   Power is the rate of doing work. In other words, it is the work done per second. If $P$ is the power, then power is expressed as:

$$P = \frac{\text{force} \times \text{distance}}{\text{time}} = \text{force} \times \frac{\text{distance}}{\text{time}} = \text{force} \times \text{velocity} = Fv \qquad (11.6)$$

**Energy:**   Energy is the capacity to do work. It exists in many forms like kinetic, potential, electrical, heat, light, etc. In engineering mechanics, both kinetic and potential energies are important. These two energies are combined to call mechanical energy. Energy possessed by a body by virtue of its motion or velocity is called *kinetic energy* (KE). If a body of mass $m$ moves from rest (i.e. initial velocity $u = 0$) with linear velocity $v$, then kinetic energy due to linear velocity is:

$$KE = \frac{1}{2}mv^2 \qquad (11.7)$$

KE of rotating body similarly expressed in terms, moment of inertia $I$, angular velocity $\omega$ as:

$$KE = \frac{1}{2}I\omega^2 \qquad (11.8)$$

Thus total KE energy of a body is:

$$\text{Total KE} = \frac{1}{2}mv^2 + \frac{1}{2}I\omega^2 \qquad (11.9)$$

Potential energy (PE) or position energy or datum energy is the energy possessed by body of weight $W$ by virtue of its position with respect to a particular datum. If body of weight $W$ is at a height of $h$ above the ground, PE of the body with respect to ground as datum is written as:

$$PE = W \cdot h \qquad (11.10)$$

**Impulse:**   When a large constant force $F$ is applied over a very small time $\Delta t$, it is called *impulse*. Thus impulse = $F \cdot \Delta t$. It is a vector quantity.

**Torque:**   Torque is the turning moment of a force on a body on which it acts. Let a tangential force $F$ act on a rotating circular body of radius $r$ at the circumference. Under the action of this force, the body moves at a small angle $\theta$, the distance moved by the force is equal to arc of the small angle $\theta$ and this distance is equal to $r\theta$ (Figure 11.1).

Now the work done (WD) by force = $F \times r\theta$ = $(F \times r) \times \theta$.

But $(F \times r)$ is equal to torque $T$.

Hence WD by this torque = $T\theta$. Here $\theta$ is in radians.

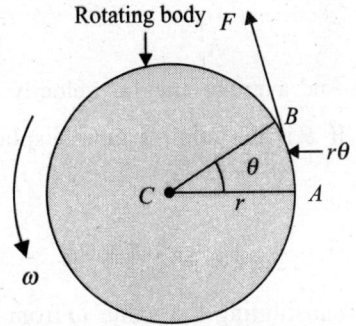

**Figure 11.1**   Torque on rotating body.

**Angular motion:**   Some bodies like pulley, shaft, flywheel, etc. have rotational or angular motion which takes place about the geometric axis of the body.

**Angular velocity:**   It is the rate of change of angular displacement of a body and is expressed in radians/second.

Let the rotating body, shown in Figure 11.1, rotate with angular velocity which is denoted by $\omega$. The angular displacement of a body is $\theta$ in time $t$.

Hence angular velocity is given by:

$$\omega = \frac{\theta}{t} \tag{11.11}$$

It is mathematically written as:

$$\omega = \frac{d\theta}{dt} \tag{11.12}$$

If the body rotates $N$ revolutions per minute (rpm), then:

$$\omega = \frac{2\pi N}{60} \, \text{rad/sec} \tag{11.13}$$

The relationship of angular and linear velocity may also be established. Let a body undergo a linear displacement in a circular path of radius $r$ in time $t$ be equal to the arc $r\theta$. Then the linear velocity is:

$$v = \frac{r\theta}{t} = r \times \frac{\theta}{t}$$

$$\therefore \qquad v = r\omega \tag{11.14}$$

**Angular acceleration:**   It is the rate of change of angular velocity which is expressed in radians/sec$^2$ and is denoted $\alpha$.

To find an expression of angular acceleration, consider the initial and final angular velocity of the body in Figure 11.1 to be $\omega_0$ and $\omega$. Let the time taken by the body to change its velocity from $\omega_0$ to $\omega$ be $t$ seconds. Since in time $t$ seconds, the angular velocity of the body has increased (or decreased) steadily from $\omega_0$ to $\omega$ at the rate of acceleration $\alpha$ radians/sec, therefore,

$$\omega = \omega_0 + \alpha t \qquad (11.15)$$

And average angular velocity is $\dfrac{\omega_0 + \omega}{2}$

If $\theta$ is the total angular displacement in time $t$ seconds

$$\theta = \text{Average velocity} \times t$$

or

$$\theta = \left(\frac{\omega_0 + \omega}{2}\right) \times t \qquad (11.16)$$

Substituting the value $\omega$ from Eq. (11.16)

$$\theta = \left(\frac{\omega_0 + (\omega_0 + \alpha t)}{2}\right) \times t$$

or

$$\theta = \left[\omega_0 t + \frac{1}{2}\alpha t^2\right] \text{ radians} \qquad (11.17)$$

From Eq. (11.16),

$$t = \frac{\omega - \omega_0}{\alpha}$$

Substituting this value in Eq. (11.16),

$$\theta = \left(\frac{\omega_0 + \omega}{2}\right) \times \left(\frac{\omega - \omega_0}{\alpha}\right) = \frac{\omega^2 - \omega_0^2}{2\alpha}$$

From the above expression, we get,

$$\omega^2 = \omega_0 + 2\alpha\theta \qquad (11.18)$$

and

$$\alpha = \frac{\omega^2 - \omega_0^2}{2\theta} \qquad (11.19)$$

**Angular displacement:**  It is the total distance through which a body rotates. If a body rotates with constant angular velocity $\omega$, then in time $t$ seconds, the angular displacement is $\omega t$.

**Centripetal force:**  When a body moves in a circular path with constant velocity, it suffers a continuous change in its direction and velocity at every point of its motion. According to the Newton's first law of motion, an external force is acting continuously on the body to produce a change in the direction of the body. The body tends to move along the tangent at every point of its motion due to inertia. Therefore, this external force must be at the right angles to the direction of motion of the body. The force which changes its direction, acts along the radius towards the centre at every point of the circular path. This force is called *centripetal force*.

**Centrifugal force:**  According to Newton's third law, the force which acts opposite to the centripetal force is called *centrifugal force*. Thus the centrifugal force always tends the body to throw the body away from the centre of circular path.

**Centripetal acceleration:**  To obtain an equation of centripetal acceleration, consider Figure 11.2 in which a particle which is moving with initial linear velocity $v$ along a circular path of radius $r$. At time $dt$, the particle moves from $A$ to $B$. When the particle was at $A$, its velocity was along tangential direction $AE$. Similarly at $B$, the direction its velocity $v_f$ along $BD$.

Now consider Figure 11.3 in which $ae$ represents the initial velocity $v_i$ in direction and magnitude at $A$ in some scale and $ad$ represents the magnitude and direction of velocity $v_f$ at $B$ in the same scale. Join $ed$. In the velocity triangle $ade$, $ae$ represents the initial velocity, $ad$ represents the final velocity and hence $de$ represents the change in velocity in time $\delta t$.

$$\text{Now acceleration} = f = \frac{\text{change in velocity}}{\text{time } dt} = \frac{de}{dt} \qquad (11.20)$$

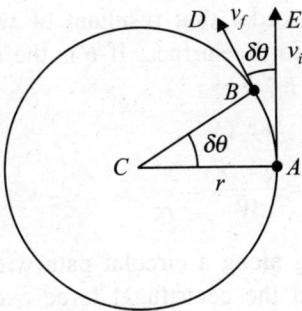

**Figure 11.2**  Centripetal acceleration.      **Figure 11.3**  Velocity triangle.

The time interval $dt$ is considered to be very small, therefore, chord length $AB$ is equal to be assumed to be arc length $AB$. Thus triangle $ACB$ and triangle $ead$ are similar. Hence:

$$\frac{de}{AB} = \frac{ae}{CA} \qquad (11.21)$$

But $AB = v_i \times dt$, $ae = v_i$ and $OA = r$
Hence Eq. (11.21) can be written as:

$$\frac{de}{v_i \times dt} = \frac{v_i}{r}$$

$$\therefore \qquad \frac{de}{dt} = \frac{v_i^2}{r} \qquad (11.22)$$

Equating (11.21) and (11.22), we get acceleration

$$f = \frac{v_i^2}{r}$$

But by Eq. (11.14)

$$v = r\omega$$

Hence Eq. (11.22) becomes

$$f = \frac{r^2\omega^2}{r} = r\omega^2 \text{ rad/sec}^2$$

When a body of mass $m$ moves with a constant velocity $v$ in a circular path of radius $r$, the centrifugal force induced by the body is equal to mass multiplied by the acceleration.
Hence centrifugal force (CF) is:

$$\text{Centrifugal force} = m\frac{v^2}{r} = mr\omega^2 \qquad (11.23)$$

Centripetal force by definition is equal to centrifugal force.

When a vehicle moves along a highway curve or train along a railway curved track, due to centrifugal force, the load on the two wheels will not be equal; the outer wheel is subjected to more load. To avoid this unbalanced load on the wheels, it is necessary to raise the pavement towards the outer side. This raising of outer edge of a curved road or railway is called *superelevation* or *cant*. The rise of outer is made in order that resultant of weight of the vehicle and centrifugal force should be perpendicular to the surface. If $\theta$ is the angle of rise then:

$$\tan\theta = \frac{\text{Centrifugal force}}{\text{Weight of the vehicle}} = \frac{\dfrac{W}{g}\cdot\dfrac{v^2}{r}}{W} = \frac{v^2}{rg} \qquad (11.24)$$

**EXAMPLE 11.1**  A body weighing 1000 N is moving along a circular path with a velocity of 15 m/sec. If the radius of the path is 400 m, find the centrifugal force exerted on the wheels.

**Solution**    Using Eq. (11.23)

$$\text{CF} = m\frac{v^2}{r} = \frac{1000}{g}\cdot\frac{15^2}{400} = \frac{1000\times15\times15}{9.81\times400}$$

or                      CF = 57.339 N      (Answer)

**EXAMPLE 11.2**  An aircraft is flying with a speed of 180 km/h in a curve of 150 m. Find the angle at which it must be banked with the horizontal.

**Solution**    Speed of the aircraft $v = 180$ km/h

or

$$v = \frac{180\times1000}{60\times60} = 50 \text{ m/sec}$$

Using Eq. (11.24)

$$\tan\theta = \frac{v^2}{rg} = \frac{50^2}{150\times9.81} = 169894$$

Hence

$$\theta = 59°31.13'      \text{(Answer)}$$

**Virtual work:**  The work done by a force on a body due to small virtual or imaginary displacement of the body is called *virtual work*.

## 11.4  MOTION OF A BODY UNDER CONSTANT ACCELERATION

Let a particle move along a straight path with initial velocity $u$. After time $t$ seconds, its velocity increases to $v$ with constant acceleration of $f$. Then final velocity becomes $v = u + ft$. Average velocity of the particle is $(u + v)/2$. If $s$ is the distance moved by the parcel in time $t$, then:

$$s = \text{average velocity} \times \text{time } t = \left(\frac{u + v}{2}\right) \times t = \left(\frac{u + u + ft}{2}\right) \times t = ut + \frac{1}{2}ft^2 \quad (11.25)$$

Again from $v = u + ft$, we get $t = \dfrac{v - u}{f}$. Hence Eq. (11.25) can also be expressed as:

$$s = \left(\frac{u + v}{2}\right) \times \left(\frac{v - u}{f}\right) = \frac{v^2 - u^2}{2f}$$

or $\qquad\qquad\qquad\qquad 2fs = v^2 - u^2 \qquad\qquad\qquad\qquad\qquad\qquad (11.26)$

or $\qquad\qquad\qquad\qquad v^2 = u^2 + 2fs$

Now if a particle moves down to earth surface under gravity,

Then $v = u + gt$, $s = ut + \dfrac{1}{2}gt^2$, $v^2 = u^2 + 2gs$. The value of acceleration $g$ due to gravity is 9.81 m/sec$^2$. If the particle moves upwards against gravity, the value of $g$ becomes negative.

***EXAMPLE 11.3***  Two cars start off to race with in velocities $u_1$ and $u_2$ and travel in a straight path with uniform accelerations $f_1$ and $f_2$ respectively. If the race ends in a dead heat, prove that the length of the race $s$ is:

$$s = \frac{2(u_1 - u_2)(u_1 f_{21} - u_2 f_1)}{(f_2 - f_1)}$$

***Solution***  If $t$ is the time for which the cars run
Using the relation

$$s = ut + \frac{1}{2}ft^2 \qquad\qquad\qquad\qquad\qquad\qquad (i)$$

$$s_1 = u_1 t + \frac{1}{2}f_1 t^2$$

and $\qquad\qquad\qquad\qquad s_2 = u_2 t + \dfrac{1}{2}f_2 t^2$

Since both the cars cover the same distance, $s_1 = s_2$

or $\qquad\qquad\qquad u_1 t + \dfrac{1}{2}f_1 t^2 = u_2 t + \dfrac{1}{2}f_2 t^2$

or $\qquad\qquad\qquad u_1 t - u_2 t = \dfrac{1}{2}f_2 t^2 - f_1 t^2 = \dfrac{1}{2}t^2(f_2 - f_1)$

or $\qquad\qquad\qquad t(u_1 - u_2) = \dfrac{1}{2}t^2(f_2 - f_1)$

$$\therefore \qquad t = \frac{2(u_1 - u_2)}{(f_2 - f_1)} \qquad \text{(ii)}$$

Now substitute this value if $t$ from Eq. (ii) in (i),

$$s = u \times \frac{2(u_1 - u_2)}{(f_2 - f_1)} + \frac{1}{2} f_1 \frac{4(u_1 - u_2)^2}{(f_2 - f_1)^2}$$

Simplifying further,

$$s = \frac{2(u_1 - u_2)(u_1 f_2 - u_2 f_1)}{(f_2 - f_1)} \qquad \text{Shown}$$

***EXAMPLE 11.4*** A motorist travelling at a speed of 86.4 km/h suddenly sees a man 100 m ahead. He instantly slows the engine and applies the brake, so as to stop the car 10 m ahead of the man. Calculate the time required by the motorist to stop the car.

***Solution*** In this example initial velocity $u = 86.4$ km/h $= \dfrac{86.4 \times 1000}{60 \times 60} = 24$ m/sec, the final velocity $v = 0$,

Let the distance to be covered by the motorist be $s$ and $s = 100 - 10 = 90$ m
Let $f$ be acceleration (retardation) of the motorist.
Using the relation

$$v^2 = u^2 + 2gs$$

or
$$0 = 24^2 + 2 \times f \times 90$$

$$\therefore \qquad f = -\frac{24^2}{2 \times 90} = -3.2 \text{ m/sec}^2$$

Minus sign indicates retardation.
Using the relation

$$v = u + ft$$

or
$$0 = 24 + (-3.2) \times t$$

$$\therefore \qquad t = \frac{24}{3.2} = 7.5 \text{ seconds} \qquad \text{(Answer)}$$

## 11.5 VIRTUAL WORK

It has been already defined in Section 11.3 that the work done is the product of force on a body and its displacement in the direction of force. In Figure 11.4, force $F$ acts on the body at an angle of $\theta$ to the direction of the body. Hence the force in the direction of the body is $F \cos \theta$ and work done by the body is $F \cos \theta \times S$. If $\theta$ is 90°, no work is done and if $\theta$ is >90° and <180° work done is negative, i.e. body is displaced in opposite direction.

If a body is in equilibrium under the action of system of forces, the work done is zero. If it is assumed that a body in equilibrium undergoes an infinite small

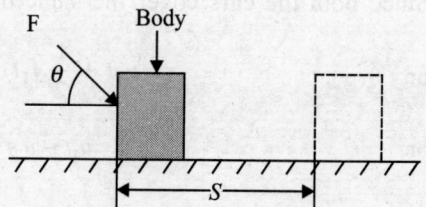

**Figure 11.4** Work done.

imaginary displacement, some work is imagined to be done. Such imaginary work is called *virtual work*. This concept of virtual work is useful to determine the unknown forces in beams, lifting machines, ladders and some structures.

## 11.5.1   Application of Virtual Work in Beam

This application is shown with the help of the following numerical examples.

**EXAMPLE 11.5**   Using the principle of virtual work, determine the reactions of a beam *AB* of span 10 m. The beam carries a point load of 3 N at *C* which is at a distance of 4 m from hinged point *A*.

*Solution*   Span of the beam, point load at *C* are shown in Figure 11.5.

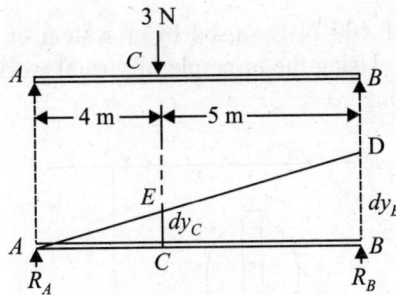

**Figure 11.5**   Application of virtual work concept in beam.

Let the beam has the virtual displacement $dy_B$ at *B* in upward direction and at *C* the virtual displacement $dy_C$ in the downward direction due to reaction $R_B$ and point load 3 N respectively. Reaction $R_A$ has zero displacement as the point *A* is hinged.

From the similar triangles *ABD* and *ACE*,

$$\frac{AB}{AC} = \frac{BD}{CE}$$

or

$$\frac{10}{4} = \frac{dy_B}{dy_C}$$

∴

$$dy_C = \frac{4}{10} dy_B \tag{i}$$

Algebraic sum of virtual WD by all the forces = Virtual WD by $R_B$ + Virtual WD by point load 3 N + virtual WD by reaction $R_A$
Sum of virtual WD = 0
Hence

$$0 = R_B \times dy_B - 3 \times dy_C + 0$$

Substituting the relationship from Eq. (i)

$$0 = R_B \times dy_B - 3 \times \frac{4}{10} dy_B$$

or

$$0 = dy_B(R_B - 1.2)$$

or

$$R_B - 1.2 = 0$$

$$R_B = 1.2 \text{ N} \quad \text{(Answer)}$$

and

$$R_A = (3 - 1.2) = 1.8 \text{ N} \quad \text{(Answer)}$$

## 11.5.2 Application of Virtual Work in Lifting Machine

Similarly principle of virtual work can be used in lifting machines. The following numerical example is given for lifting machine.

**EXAMPLE 11.6** A weight of 400 N is raised by a system of frictionless pulleys of equal radius as shown in Figure 11.6. Using the principle of virtual work, find the effort $P$ which can keep the weight in equilibrium.

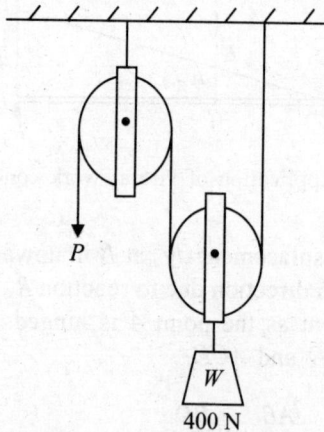

**Figure 11.6** Application of virtual work in lifting machine.

**Solution** Assume virtual displacement of weight 400 N to be $y$
Then virtual displacement of effort $P$ will be $2y$.
Algebraic sum of virtual WD by $P$ and $W$ = Virtual WD by $P$–Virtual WD by $W$

or

$$0 = P \times 2y - W \times y$$

or

$$2P = W = 400 \text{ N}$$

∴

$$P = \frac{400}{2} = 200 \text{ N} \quad \text{(Answer)}$$

## 11.5.3 Application of Virtual Work in Framed Structures

Again application of the concept of virtual work can be applied in framed structures. The member of the structures in which the force is to be determined is assumed to be removed from the structure. The member is replaced by two forces each equal to $F$ acting at the ends of the member. By applying the principle of virtual work, force in the member is determined. A numerical example is given to show its application.

**EXAMPLE 11.7**   Five rods $AB$, $BC$, $CD$, $DA$ and $BD$ each of equal length and equal cross section are pin-jointed together so as to form a plane frame $ABCD$. The frame $ABCD$ is a rhombus with one horizontal diagonal $BD$. The frame is suspended from the topmost point $A$ as shown in Figure 11.7. A load of weight $W$ is attached to the lower point $C$. Neglecting the self-weight of the frame and applying the principle of virtual work, determine the magnitude of the force in the member $BD$ in term of $W$.

**Solution**   Figure 11.7 is the visual of Example 11.7. In Figure 11.8, member $BD$ is removed with two forces applying forces $F$ at $B$ and $D$ as shown. Let $X$ is the middle point of rod $BD$.

Geometry of Figure 11.8 shows that angle $\angle BXC$ is right angle before virtual displacement. Let us assume that this angle will remain a right angle even after virtual displacement.

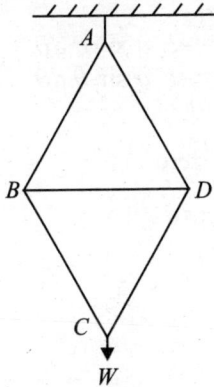

**Figure 11.7**   Frame analysis to apply the concept of virtual work.

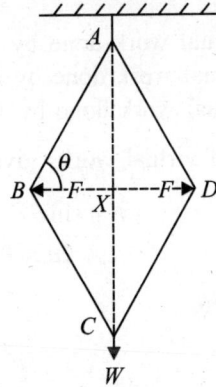

**Figure 11.8**   Member $BD$ is removed by forces $F$.

Since $ABCD$ is a rhombus,

$$AB = BC = CD = DA = a$$

From Figure 11.8,

$$AX = AB \sin = a \sin\theta$$
$$BX = AB \cos\theta = a \cos\theta$$
$$AC = AX + CX = 2a \sin\theta$$

If the frame undergoes a small virtual displacement $d\theta$ as shown in Figure 11.9, points $B$ and $D$ move inward and point $C$ moves downward keeping $A$ fixed as a result of defection.

Now virtual horizontal displacement of $B$ is given by

$$d(BX) = d(a \cos\theta) = -a \sin\theta \, d\theta$$

Similarly virtual displacement of $DX$ is equal to $-a \sin\theta \, d\theta$
Similarly virtual horizontal displacement of $C$ is given by

$$d(AC) = d(2a \sin\theta) = 2a \cos\theta \, d\theta$$

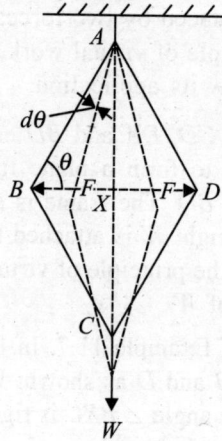

**Figure 11.9**  Frame undergoes a small displacement $d\theta$.

Virtual work done by $F$ at $B = F(-a \sin \theta\, d\theta) = -F\, a \sin \theta\, d\theta$   (i)

and  Virtual work done by $F$ at $D = F(-a \sin \theta\, d\theta) = -F\, a \sin \theta\, d\theta$   (ii)

and  Virtual work done by $W$ at $C = W \times (2a \cos \theta\, d\theta)$   (iii)

But summation of virtual works given by (i), (ii) and (iii) is zero.

$\therefore$   $$-F\, a \sin \theta\, d\theta - F\, a \sin \theta\, d\theta + W \times 2a \cos \theta\, d\theta$$

or   $$2a \times F \sin \theta\, d\theta = 2a \times W \cos \theta\, d\theta$$

Dividing by $2a\, d\theta$,

$\therefore$   $$F \sin \theta = \frac{W \cos \theta}{\sin \theta} = \frac{W}{\tan \theta}$$

But the value of $\theta = 60°$ $\because$ $ABD$ is equilateral
Hence

$$F = \frac{W}{\tan 60°} = \frac{W}{\sqrt{3}} \qquad \text{(Answer)}$$

## 11.5.4  Application of Virtual Work in Ladder

Similarly again application of the concept of virtual work can be applied in analyzing forces of friction in ladder. The following is an example of concept of virtual work done in ladder.

**EXAMPLE 11.8**  A ladder of weight 250 N and length $a$, rests against a smooth vertical wall and a rough horizontal floor making an angle of 45° with the horizontal as shown in Figure 11.10. Using the method of virtual work, find the force of friction of the floor.

**Solution**  Figure 11.10 shows the ladder leaning against a smooth wall $OB$ at an angle $\theta$, its self-weight 250 N acting downwards through its middle point $G$, reactions $R_A$, $R_B$, force of friction $F_A$ against a rough floor and length of ladder $a$. Since the wall is smooth, no reaction for friction exists at $B$.

From Figure 11.10,

$$y_G = AG \sin \theta = (a/2) \sin \theta \text{ and } x_A = AB \cos \theta = a \cos \theta$$

In Figure 11.11, a small virtual displacement is given to the ladder so that the point $A$ moves to point $A'$ with the displacement $dx_A$ keeping the length of the ladder fixed. As a result of this movement, ladder will move downwards through a distance $dy_B$.

Now considering the virtual displacement of point $G$:

$$dy_G = d\left(\frac{a}{2} \sin \theta\right) = \frac{a}{2} \cos \theta \, d\theta$$

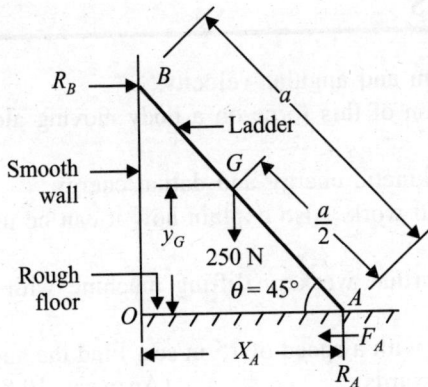

**Figure 11.10**  Ladder resting on rough floor.

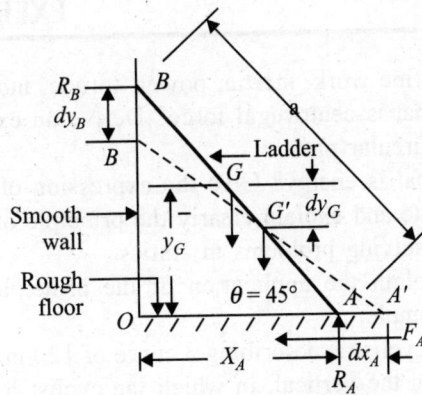

**Figure 11.11**  Ladder under small virtual displacement.

Similarly the virtual horizontal displacement is:

$$dx_A = d(\cos \theta) = -a \sin \theta \, d\theta$$

Virtual work done by reaction $R_A = R_A \times$ virtual displacement upwards $= R_A \times 0 = 0$

Virtual work done by reaction $R_B = R_B \times$ virtual displacement horizontally $= R_B \times 0 = 0$

Virtual work done by force of friction at $A = F_A \times dx_A = F_A \times (-a \sin \theta \, d\theta) = -F_A \, a \sin \theta \, d\theta$

Virtual work done by the weight of the ladder $= W \times dy_G$

$$= 250 \times \frac{a}{2} \cos \theta \, d\theta$$

Total virtual work done $= 0$

or

$$250 \times \frac{a}{2} \cos \theta \, d\theta - F_A \, a \sin \theta \, d\theta = 0$$

or

$$F_A \sin \theta = 125 \cos \theta$$

$$\therefore \qquad F_A = \frac{125 \cos \theta}{\sin \theta} = \frac{125}{\tan \theta} = \frac{125}{\tan 45°} = 125 \text{ N} \qquad \text{(Answer)}$$

## 11.6 CONCLUSION

Definitions of common terms related to kinetics of rigid body are given at the beginning. Newton's laws are defined. Motions of bodies under constant acceleration, kinetic energies on linear and rotational motion are presented. Centrifugal and centripetal forces are defined. Their importance of moving bodies in curve path, a few numerical problems are also solved. Virtual work is defined and its application in different situations like beam, framed structures, lifting machines, ladder is also demonstrated. A few numerical problems are solved to show its applications.

## EXERCISES

11.1 Define work, inertia, power, torque, momentum and angular velocity.

11.2 What is centrifugal force? Derive an expression of this force on a body moving along a circular path.

11.3 What is energy? Give the expression of total kinetic energy and datum energy.

11.4 State and explain clearly the principle of virtual work. Also explain how it can be used in solving problems in statics.

11.5 Explain the application of the principle of virtual work on lifting machine with an example.

11.6 A cyclist is describing a curve of 120 m radius with a speed of 15 m/sec. Find the angle with the vertical, in which the cyclist leans inwards.       (Answer   10.83°)

11.7 A weight of 100 N is raised by two frictionless pulleys of same diameter as shown in Figure 11.12. Determine the effort $P$ required to hold the weight in equilibrium, using the principle of virtual work.

11.8 A ladder of weight 600 N and length $a$ rests against a smooth vertical wall and a rough horizontal floor making an angle of 45° with the horizontal as shown in Figure 11.13. Using the method of virtual work, find the force of friction of the floor.

(Answer   300 N)

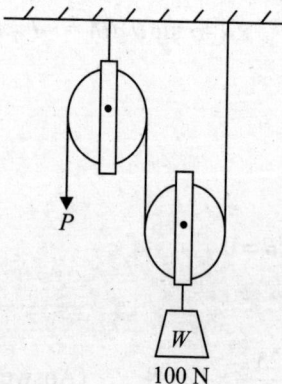

**Figure 11.12**   Visual of Exercise 11.7.

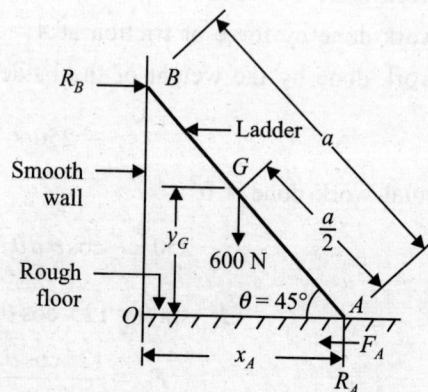

**Figure 11.13**   Ladder resting on rough floor.

**11.9** Using the principle of virtual work, determine the reactions of a beam $AB$ of span 5 m shown in Figure 11.14. The beam carries a point load of 4 N at $C$ which is at a distance of 2 m from hinged point $A$. (Answer $R_A = 1.6$ N, $R_B = 2.4$ N)

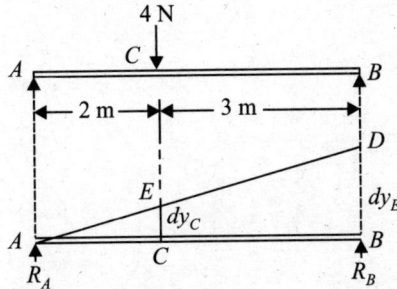

**Figure 11.14** Visual of Exercise 11.9.

# Part II

## Strength of Materials

# CHAPTER 12

# Simple Stress and Strain

## 12.1 INTRODUCTION

Materials may be classified into solid, plastic and elastic. A design engineer should have proper knowledge of strength of these materials to resist the deformation when external forces act on it otherwise safe and sound design of any structure required in our day-to-day life is not possible. When an external force is applied to the above materials, deformation of the body takes place more or less depending upon the type of the body within their elastic limit. This deformation also depends on type of external force, type of application and nature of the structure or body. The force of resistance offered by the body or structure is due to the cohesive force between the molecules of the body. The resistance by which the material of the body opposes the deformation, is known as *strength of materials*. The deformation of rigid body is quite small or negligible compared to elastic and plastic bodies. For example, the deformation of elastic body is quite visible under the action of external forces and it returns to its original position after the force is removed although some residual deformation may remain in it. The plastic body also goes on deforming under the action of external force and this deformation is almost permanent. The material does not regain its original dimensions after the removal of the force. Thus in practice, all bodies will be treated to be deformable within their elastic limit. The body tends to offer the force of resistance by virtue of its strength to prevent the deformation. When the body or structural member is unable to offer the necessary resistance against the external forces, the deformation goes on increasing, leading to the failure of the member or body. Structural members are always under the action of external forces and hence they continue to offer resistance against deformation. For safety of the structure, the force of resistance developed should always be within their elastic limit. Stress and strain that are developed in the members are the most important parameters required to analyze the safety of the structure.

## 12.2 CONCEPT OF STRESS

The study of the strength of materials starts with the concept of stress produced by the applied loads on a structure or a machine and the members that make up the system. The applied loads initially will be simple direct axial loads, direct shearing forces and bearing loads. Stress is defined as load per unit area.

## 12.2.1   Compressive Stress

Consider Figure 12.1(a), in which the load $P$ acts on a bar $MN$ of cross-sectional area $A$. The bar develops a resisting force $R$ under the action of applied compressive forces. The applied load $P$ is assumed to be uniformly distributed over the entire cross section. To determine the resisting force $R$, it is required to imagine a section plane $xx$ as shown, and equilibrium of any one part may be considered. Each part will be in equilibrium under the action of applied and resisting forces. With reference to Figure 12.1(a), consider the equilibrium of left-hand part shown in Figure 12.1(b). This force $R$ is equal to $P$. Thus the average intensity of pressure or stress is expressed mathematically as:

$$\sigma = \frac{R}{A} = \frac{P}{A} \qquad (12.1)$$

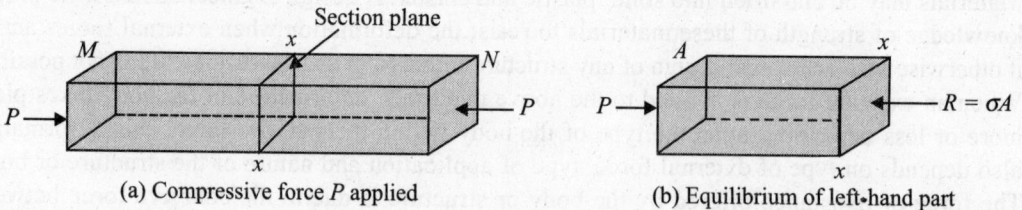

(a) Compressive force $P$ applied        (b) Equilibrium of left-hand part

**Figure 12.1**   Compressive stress.

If the stress is not uniform over a section, the intensity of stress at any point is expressed on an infinitesimal area $dA$ over which the load $dP$ is applied. The stress at a point can be expressed in differential form as:

$$\sigma = \frac{dP}{dA} \qquad (12.2)$$

This stress is developed under direct compressive load and this stress is called *direct compressive stress*. This compressive force has a tendency to shorten the member $MN$.

## 12.2.2   Tensile Stress

Now consider Figure 12.2(a) in which the bar $MN$ is subjected to a pull $P$. For equilibrium end forces $P$ must be equal and opposite. Under such forces, the deformable bar $MN$ has a tendency to increase its length. The resisting forces inside the bar are shown in Figure 12.2(b). The uniform stress developed in this situation is given by same Eq. (12.1). As the bar has the tendency to increase its length, the stress is called *tensile stress*.

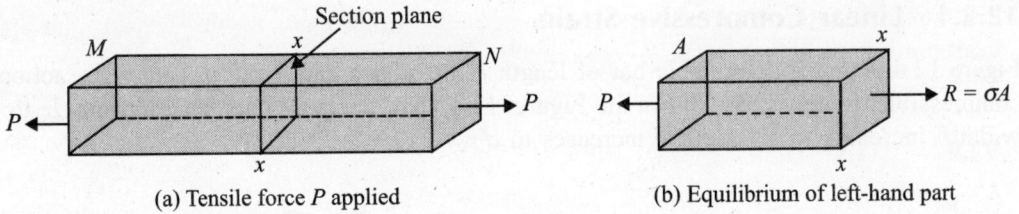

(a) Tensile force $P$ applied     (b) Equilibrium of left-hand part

**Figure 12.2**   Tensile stress.

### 12.2.3  Shear Stress

The stress induced in a body when two equal and opposite forces are acting tangentially across the resisting section, as a result of which the body tends to shear off across the section. This stress is known as *shear stress*. Such stress occurs due to force acting parallel to the cross section of the member. It acts tangential to the cross section and opposes the angular deformation due to applied forces. Shear stress is also called *tangential stress*.

Consider a rectangular block $BCDE$ in Figure 12.3. The bottom of this block is fixed on a surface $GH$. Let a force $P$ be applied tangentially along the top surface of the block as shown in Figure 12.3. This force is called *tangential force*. For equilibrium of the block, the surface $GH$ offers a tangential reaction $R$ which is equal to $P$. Intensity of this force per unit area of the surface is called *shear* or *tangential stress* $\tau$. If $A$ is the area of the surface on which this tangential force acts, then shear stress $\tau$ is equal to:

$$\tau = \frac{R}{A} = \frac{P}{A} \tag{12.3}$$

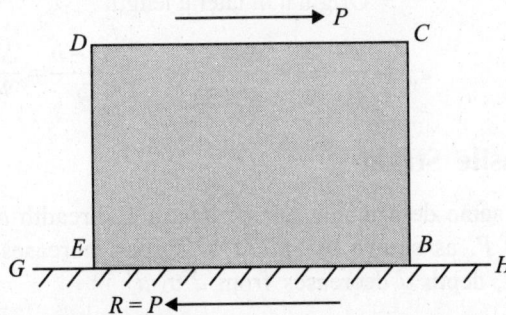

**Figure 12.3**   A fixed rectangular block with tangential force $P$ applied at top surface.

### 12.3  CONCEPT OF STRAIN

Strain is a measure of deformation of the body under the action of external forces. It is non-dimensional as it is expressed in the ratio of increase or decrease in length to original length. There are three types of strain.

## 12.3.1   Linear Compressive Strain

Figure 12.4 shows a deformable bar of length $L$, breadth $b$ and depth $d$. Under the action of compressible forces $P$ as shown in Figure 12.4, bar decreases its length from $L$ to $L'$, width $b$ increases to $b'$, depth $d$ increases to $d'$.

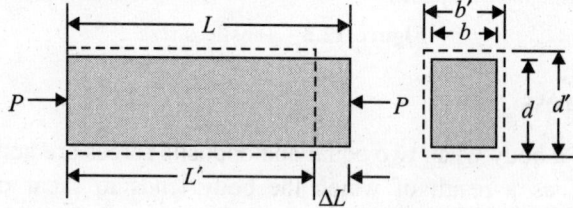

**Figure 12.4**   Deformation under compressive force.

Linear compressive strain is:

$$e_x = \frac{\text{Decrease in length}}{\text{Original length}}$$

or

$$e_x = \frac{\Delta L}{L} \tag{12.4}$$

Linear lateral (tensile) strain is:

$$e_y = \frac{\text{Change in lateral length}}{\text{Original in lateral length}}$$

$$e_y = \frac{d' - d}{d} = \frac{\Delta d}{d} \quad \text{or} \quad \frac{b' - b}{b} = \frac{\Delta b}{b} \tag{12.5}$$

## 12.3.2   Linear Tensile Strain

Figure 12.5 shows the same deformable bar of length $L$, breadth $b$ and depth $d$. Under the action of tensile forces $P$, as shown in Figure 12.5, bar increases its length from $L$ to $L'$, width $b$ decreases to $b'$, depth $d$ decreases from $d$ to $d'$.

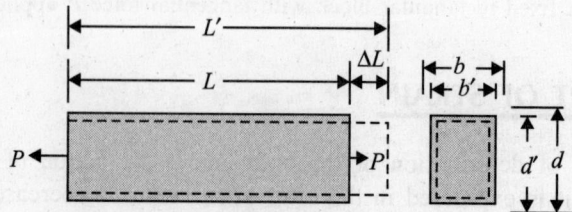

**Figure 12.5**   Deformation under tensile force.

Linear tensile strain is:

$$e_x = \frac{\text{Increase in length}}{\text{Original length}}$$

or

$$e_x = \frac{\Delta L}{L}$$ (using Eq. (12.4))

Linear lateral (compressive) strain is:

$$e_y = \frac{\text{Change in lateral length}}{\text{Original in lateral length}}$$

$$e_y = \frac{d - d'}{d} = \frac{\Delta d}{d} \quad \text{or} \quad \frac{b - b'}{b} = \frac{\Delta b}{b}$$ (using Eq. (12.5))

### 12.3.3 Shear Strain

Figure 12.6 shows a rectangular bar subjected to shear forces $P$ on its top and bottom faces. If the block does not fail in shear, a shear deformation occurs as shown in Figure 12.6. Again if the bottom face of the block is fixed, block has deformed to the position $AB'C'D$. In other words, it may be said that the face $BCDA$ has been distorted to the position $B'C'DA$ through angle $CD\ C' = \phi$. Let the horizontal displacement of upper face of the block be $d\ell$ and height be $\ell$.

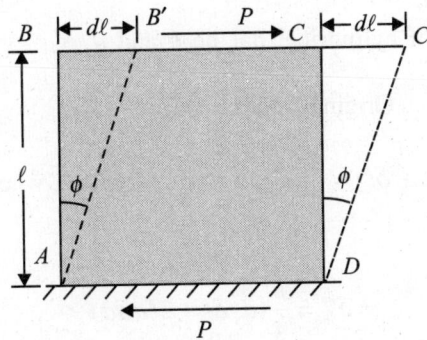

**Figure 12.6** Deformation under shear force.

Now shear strain $= \dfrac{\text{Transverse displacement}}{\text{Distance from the lower face}} = \dfrac{d\ell}{\ell}$ (12.6)

Thus shear strain $= \dfrac{d\ell}{\ell} = \tan \phi$ (12.7)

Since $\phi$ is very small, $\phi = \tan \phi = \dfrac{d\ell}{\ell} =$ shear strain (12.8)

Hence the angular deformation $\phi$ in radian measure represents the shear strain.

## 12.3.4 Volumetric Strain

If a volume $V$ of solid or fluid is under identical shear stresses $\tau$ from all directions, change of volume $\Delta_v$ takes place. In such situation volumetric strain may be defined as:

$$e_v = \frac{\text{Change of volume}}{\text{Original volume}} = \frac{\Delta_v}{V} \qquad (12.9)$$

**When the bar is rectangular:**

Let the length, width and depth of bar of length be $\ell$, $b$ and $d$ respectively.

$$\text{Then original volume } V = \ell \times b \times d = \ell b d$$

Let the increase of length, breadth and depth after application of stress be $\delta\ell$, $\delta b$ and $\delta d$. Then the final volume will be $(\ell + \delta\ell) \times (b + \delta b) \times (d + \delta d) = \ell b d + b d \delta\ell + \ell b \delta d + \ell d \delta b$ (ignoring the products of smaller terms).

Change in volume

$$\delta V = b d \delta\ell + \ell b \,\delta d + \ell d \delta b$$

Volumetric strain

$$e_v = \frac{\text{Change in volume}}{\text{Original volume}} = \frac{b d \delta\ell + \ell b \delta d + \ell d \delta b}{\ell b d} = \frac{\delta\ell}{\ell} + \frac{\delta d}{d} + \frac{\delta b}{b} = e_\ell + e_d + e_b$$

$$\therefore \qquad \text{Volumetric strain} = \text{Strain in length} + \text{Strain in depth} + \text{Strain in breadth} \quad (12.10)$$

**When the bar is cylindrical:**

Let the length and diameter of cylindrical bar be $\ell$ and $d$.

$$\text{Original volume } = V = \frac{\pi}{4} d^2 \ell$$

$$\text{Final volume} = \frac{\pi}{4}(d + \delta d)^2 \, (\ell + \delta\ell) = \frac{\pi}{4}(d^2 \ell + d^2 \delta\ell + 2\ell d \delta d), \text{ when higher powers of smaller terms are neglected.}$$

Change in volume

$$= \delta V = \frac{\pi}{4}(d^2 \delta\ell + 2\ell d \delta d)$$

Volumetric strain

$$= e_v = \frac{\text{Change in volume}}{\text{Original volume}} = \frac{\delta V}{V} = \frac{\frac{\pi}{4}(d^2 \delta\ell + 2\ell d \delta d)}{\frac{\pi}{4} d^2 \ell} = \frac{d^2 \delta\ell + 2\ell d \delta d}{d^2 \ell}$$

or

$$e_v = \frac{\delta\ell}{\ell} + 2\frac{\delta d}{d} \qquad (12.11)$$

In other words, volumetric strain = strain of the length + 2 times the strain of the diameter.

**When it is a sphere:**

Let the diameter $d$ of the sphere increases to $(d + \delta d)$.

$$\text{Change in volume } \delta V = \text{Original volume} - \text{Final volume}$$

or $$\delta V = \frac{\pi(d+\delta d)^3}{6} - \frac{\pi d^3}{6} = \frac{\pi d^3}{6} + + \frac{\pi \delta d^3}{6} + \frac{\pi 3 d^2 \delta d}{6} + \frac{\pi 3 d \delta d^2}{6} - \frac{\pi d^3}{6}$$

Neglecting terms with smaller values like $\delta d^3$, $\delta d^2$,

$$\delta V = \frac{\pi}{6}(3d^2 \delta d)$$

$$\text{Volumetric strain, } e_v = \frac{\text{Change in volume}}{\text{Original volume}} = \frac{\delta V}{V} = \frac{3d^2 \delta d}{d^3} = 3\frac{\delta d}{d} = 3e_d \qquad (12.12)$$

## 12.4   ELASTIC LIMIT

A material is said to be elastic when it deforms under loads and returns to its original position after removal of the applied forces. Axial loading on a member produces resistance or stress. after the applied forces are removed, stress and deformation will vanish. This property in materials exists so far the loading and deformation are kept within certain limit. The limit of intensity of stress in a material at which it has the property to regain its original shape is called *elastic limit*. If the loading is more and stress developed in the material exceeds the elastic limit, it loses to some extent its property of elasticity. It will not regain its original shape and thus it attains certain residual strain in it.

## 12.5   HOOKE'S LAW

It was Robert Hooke (1635–1730), a professor in Gresham College (UK), who for the first time investigated the behaviour of several materials subjected to gradually increasing load. After a lot of investigations, he established the relationship between stress and strain in materials. He concluded that stress developed in a material is proportional to strain within elastic limit. This relationship developed by him in 1678 between stress and strain eventually named as Hooke's Law in honour of his pioneering work.
Thus

$$\text{stress} \propto \text{strain}$$
or $$\text{stress} = \text{constant} \times \text{strain}$$
or $$\sigma = \text{constant} \times e \qquad (12.13)$$

## 12.6   ELASTIC CONSTANTS

There are four elastic constants. They are as follows:

(i) Young's modulus or modulus of elasticity ($E$)
(ii) Poisson's ratio ($\mu$)

(iii) Modulus of rigidity ($G$)

(iv) Bulk modulus ($K$)

### 12.6.1 Young's Modulus of Elasticity

It was an English scientist, Thomas Young (1773–1829) who evaluated the constant of proportionality of Hook's Law and that constant for particular material is given the name, *Young's modulus of elasticity* ($E$). Thus mathematical expression of Hooke's Law is written as:

$$\sigma = Ee \tag{12.14}$$

or

$$\frac{P}{A} = E\frac{\Delta L}{L}$$

or

$$\Delta L = \frac{PL}{AE} \tag{12.15}$$

Here $E$ is the Young's modulus of elasticity. Young's modulus $E$ is a property of the material and may have constant value in many of the cases. But depending on stress, it may have different values in compression and tension. It may be noted in Eq. (12.14) that $e$ is non-dimensional and hence $E$ has the same dimension of stress. In SI unit it is N/m$^2$.

### 12.6.2 Poisson's Ratio ($\mu$)

It has been observed in Sections 12.3.1 and 12.3.2 that longitudinal strain and lateral strain are opposite in nature. If longitudinal strain is compressive, lateral strain is tensile. Similarly, if longitudinal strain is tensile, lateral strain is compressive. Simeon Poisson, a French mathematician, who applied the molecular theory of structure of materials, had showed that lateral strain is proportional to linear strain within elastic limit. In other words, ratio of lateral strain to linear strain of a material is constant and that constant $\mu$ or $\nu$ is named as Poisson's ratio after this French mathematician. Mathematically,

$$\mu = \frac{\text{Lateral strain}}{\text{Linear strain}} = \frac{1}{m} = \frac{e_y}{e_x} \tag{12.16}$$

Here $e_y$ are $e_x$ given by Eqs. (12.4) and (12.5) respectively. It is obvious from Eq. (12.16) that Poisson's ratio $\mu$ is a non-dimensional quantity. Its value for different metals normally varies from 0.25 to 0.33, i.e. the value of $m$ is from 3 to 4. For steel and concrete, values of Poisson's ratio are taken as 0.3 and 0.15 respectively.

### 12.6.3 Modulus of Rigidity ($G$)

Modulus of rigidity $G$ is defined as the ratio of shear stress to shear strain. Mathematically,

$$G = \frac{\tau}{\phi} \tag{12.17}$$

Here $\tau$ is the shear stress given by Eq. (12.3), $\phi$ is the shear strain given by Eq. (12.8). In Eq. (12.17), shear strain $\phi$ is non-dimensional; hence dimension of $G$ is same as shear stress $\tau$, i.e. N/m$^2$.

### 12.6.4  Bulk Modulus (*K*)

If a volume $V$ of solid or fluid is under identical stresses $\tau$ from all directions, change of volume $\Delta_v$ takes place. In such situation volumetric strain may be defined as:

$$e_v = \frac{\Delta_v}{V} \tag{12.18}$$

Bulk modulus ($K$) can be defined as the ratio of stress to volumetric strain. Mathematically,

$$K = \frac{\tau}{e_v} \tag{12.19}$$

The dimension of $K$ is similar to the dimension of stress as volumetric strain is non-dimensional.

## 12.7   STRESS-STRAIN RELATIONSHIP

A mild steel bar is taken as a specimen to study the stress-strain relationship under a tensile test. When the tensile load is applied and increased gradually, from point $O$ to $A$ in Figure 12.7, the corresponding line to the point $A$ is called *limit of proportionality*. In this portion from $O$ to the point $A$, stress and strain maintain a straight line relationship, i.e. stress is proportional to strain. From $O$ to $A$, Hooke's Law is applicable, i.e. Eq. (12.14) is valid, i.e. $\sigma = Ee$. When the load on the specimen is extended beyond the limit of proportionality up to the condition $B$, material remains still elastic but relationship between stress and strain no longer remains linear. The stress at $B$ is called *stress in elastic limit*. Further extension of the specimen beyond elastic limit, produces a plastic deformation. From $B$ to $C$, strain increases at constant stress.

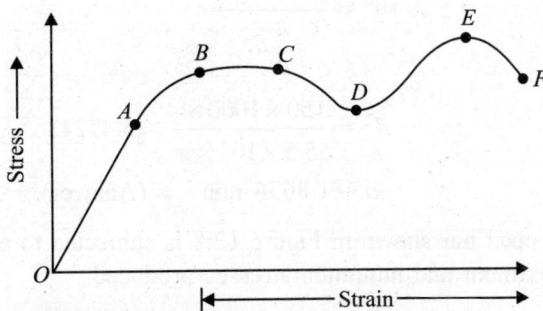

**Figure 12.7**   Stress-strain diagram of mild steel.

The stress at $C$ is called upper yield stress. At the condition shown at point $D$, material offers resistance to further extension. The stress at $D$ is called *stress in lower yield point*. Further increase of load, extension increases and a condition shown at $E$, a waist or necking of specimen is being developed. The stress at $E$ is called *ultimate tensile stress*. Further extension of specimen, load required decreases and specimen shown at $F$ breaks. The stress at $F$ is called the *stress at failure*.

**EXAMPLE 12.1** A prismatic bar has cross section 12.5 cm² and a length 2 m. The measured elongation of the bar 0.15 cm under an axial load 90 kN. Compute the tensile stress and strain in the bar.

**Solution**  Using Eq. (12.1) to find the required stress.

$$\sigma = \frac{R}{A} = \frac{P}{A} \frac{90 \times 1000}{12.5} = 7200 \text{ N/cm}^2 \quad \text{(Answer)}$$

Using Eq. (12.4) to determine the strain

$$e_x = \frac{\Delta L}{L} = \frac{0.15}{2 \times 100} = 7.5 \times 10^{-4} \quad \text{(Answer)}$$

**EXAMPLE 12.2** A steel rod of circular cross section is to carry a tensile load of 150 kN. The allowable working stress is $5.5 \times 10^4$ N/mm². Find the required diameter of the rod.

**Solution**  Here

$$\sigma = 5.5 \times 10^4 \text{ N/mm}^2$$
$$P = 150 \text{ kN}$$

The diameter of the rod = d
Area of the circular bar

$$A = \frac{\pi}{4} d^2 = \frac{\pi}{4} \times d^2$$

Applying Eq. (12.1)

$$\sigma = \frac{P}{A}$$

or

$$5.5 \times 10^4 = \frac{150 \times 1000}{\frac{\pi}{4} d^2}$$

or

$$d^2 = \frac{150 \times 1000 \times 4}{5.5 \times 10^4 \times \pi} = 3.47242$$

$$\therefore \qquad d = 1.8634 \text{ mm} \quad \text{(Answer)}$$

**EXAMPLE 12.3** A stepped bar shown in Figure 12.8 is subjected to an axially compressive load of 35 kN. Find maximum and minimum stresses produced.

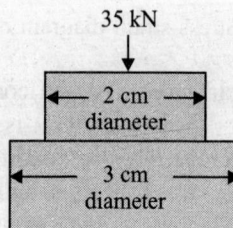

**Figure 12.8**  Visual of Example 12.3.

**Solution**   Area of the upper bar = $A_1 = \dfrac{\pi}{4}(2)^2 = 3.1415\ cm^2$

Area of the lower bar = $A_2 = \dfrac{\pi}{4}(3)^2 = 7.0685\ cm^2$

now                                 $\sigma_{max} = \dfrac{35}{3.1415} = 11.1411\ kN/cm^2$     (Answer)

and                                 $\sigma_{min} = \dfrac{35}{7.0685} = 4.9515\ kN/cm^2$     (Answer)

**EXAMPLE 12.4**   A steel wire of diameter 8 mm undergoes a pull of 4000 N. Determine the elongation of a 3.5 m long wire if Young's modulus $E = 2 \times 10^5\ N/mm^2$.

**Solution**   Given data are:

$$d = 8\ mm \text{ and hence } A = \dfrac{\pi}{4}d^2 = \dfrac{\pi}{4} \times 8^2 = 50.2654\ mm^2$$

$$P = 4000\ N$$
$$E = 2 \times 10^5\ N/mm^2$$
$$L = 3.5\ M$$

Let $\Delta L$ is the elongation of the wire of 3.5 m.
Using Eq. (12.15)

$$\Delta L = \dfrac{PL}{AE} = \dfrac{4000 \times 3.5 \times 1000}{50.2654 \times 2 \times 10^5} = 1.3926\ mm \qquad \text{(Answer)}$$

**EXAMPLE 12.5**   The safe stress, for a hollow steel column which carries an axial load of 2500 kN is 130 N/mm². The external diameter of the column is 320 mm. Determine the internal diameter.

**Solution**   External diameter of the hollow column = 320 mm
The internal diameter = $d$ mm
Given load on the column = 2500 kN
Safe stress under this load = 130 N/mm².

Sectional area required $= \dfrac{\text{Given load}}{\text{Safe stress}} = \dfrac{2500 \times 1000}{1300} = 1923.0769\ mm^2$

$\therefore$                          $\dfrac{\pi}{4}(320^2 - d^2) = 1923.0769$

or                                 $102400 - d^2 = 2448.5375$
or                                 $d^2 = 99951.4624$
$\therefore$                          $d = 316.151\ mm$     (Answer)

**EXAMPLE 12.6**   Ultimate stress for a hollow steel column which carries an axial load of 1.9 MN is 480 N/mm². If the external diameter of the column is 200 mm, determine the internal diameter. Take the factor of safety as 4.

*Solution*   Axial load on the column = 1.9 MN = $1.9 \times 10^6$ N
Ultimate stress on the column = 480 N/mm$^2$
Factor of safety = 4

Hence safe stress on the column = $\sigma = \dfrac{480}{4} = 120$ N/mm$^2$

Sectional area of the column = $A = \dfrac{1.9 \times 10^6}{120} = 15833.333$ mm$^2$

External diameter of the column = 200 mm
Let $d$ be the internal diameter

$$\therefore \qquad \frac{\pi}{4}(200^2 - d^2) = A = 15833.333$$

$$40000 - d^2 = A = 15833.333 \times \frac{4}{\pi} = 20159.625$$

or
$$d^2 = 40000 - 20159.625$$
$$\therefore \qquad d = 140.855 \text{ mm} \qquad \text{(Answer)}$$

*EXAMPLE 12.7*   A uniform steel bar of 1 m long, 5 mm diameter is subjected to a tensile force of 3000 N. Determine the tensile stress and its elongation if Young's modulus $E$ is 150 GPa.

*Solution*   Note that 1 MPa (mega Pascal) = $10^6$ N/m$^2$ and 1 GPa (giga Pascal) = $10^9$ N/m$^2$

$$\text{Tensile force } P = 3000 \text{ N}$$

$$\text{Cross-sectional area } A = \frac{\pi}{4}(5)^2 = 19.635 \text{ mm}^2$$

$$\therefore \qquad \text{Tensile stress } \sigma = \frac{P}{A} = \frac{3000}{19.635} = 152.788 \text{ N/mm}^2 \qquad \text{(Answer)}$$

Given $E = 150$ GPa = $150 \times 10^9$ N/m$^2$ = $\dfrac{150 \times 10^9}{10^6} = (150 \times 10^3)$ N/mm$^2$

If $\Delta L$ is the elongation of the 1 m long steel bar, then use Eq. (12.15)

$$\Delta L = \frac{PL}{AE} = \frac{3000 \times (1 \times 1000)}{19.635 \times (150 \times 10^3)} = 1.0185 \text{ mm} \qquad \text{(Answer)}$$

*EXAMPLE 12.8*   Mild steel is tested in a laboratory of strength of materials. The data pertaining to the laboratory test are as follows:

    (i) Length of the specimen                 = 320 mm
   (ii) Diameter of the mild steel rod      = 30 mm
 (iii) Elongation under a load of 20 kN   = 0.055 mm
 (iv) Load at yield point                   = 130 kN
  (v) Maximum load                       = 210 kN
 (vi) Length of the specimen after failure = 380 mm
(viii) Neck diameter                       = 22 mm

From the data collected from the test, determine

   (i) $E$
  (ii) yield point
 (iii) ultimate stress
 (iv) percentage (pc) of elongation
  (v) percentage (pc) of decrease in area
 (vi) safe stress adopting a factor of safety of 1.5.

*Solution*

   (i) Young's modulus $E$ can be obtained from Eq. (12.15)

$$\Delta L = \frac{PL}{AE}$$

or

$$0.055 = \frac{20 \times 1000 \times 320}{\dfrac{\pi}{4} \times 30^2 \times E}$$

Solving for $E$,

$$E = 1.6462 \times 10^5 \text{ N/mm}^2 \qquad \text{(Answer)}$$

  (ii) Stress at yield point $= \dfrac{\text{Load at yield point}}{\text{Area of specimen}} = \dfrac{130 \times 1000}{\dfrac{\pi}{4} \times 30^2} = 183.912 \text{ N/mm}^2$   (Answer)

 (iii) Ultimate stress $= \dfrac{\text{Maximum load}}{\text{Area of specimen}} = \dfrac{210 \times 1000}{\dfrac{\pi}{4} \times 30^2} = 297.089 \text{ N/mm}^2$   (Answer)

 (iv) pc of elongation $= \dfrac{\text{Increase in length}}{\text{Original length}} \times 100 = \dfrac{380 - 320}{320} \times 100 = 18.75\%$   (Answer)

  (v) pc decrease in area $= \dfrac{\text{Decrease in area}}{\text{Original area}} \times 100 = \dfrac{30^2 - 22^2}{30^2} \times 100 = 46.22\%$ (Answer)

 (vi) Safe stress $= \dfrac{\text{Yield stress}}{\text{Factor of safety}} = \dfrac{183.912}{1.5} = 122.608 \text{ N/mm}^2$   (Answer)

**EXAMPLE 12.9**  A bar consists of three lengths $L_1$, $L_2$, $L_3$ and diameter $D_1$, $D_2$, $D_3$ and is subjected to an axial pull $P$ as shown in Figure 12.9. If $E$ is the Young's modulus of elasticity, show that the total extension $\Delta L$ of the bar is given by:

$$\Delta L = \frac{4P}{\pi E} \left( \frac{L_1}{D_1^2} + \frac{L_2}{D_2^2} + \frac{L_3}{D_3^2} \right) \tag{i}$$

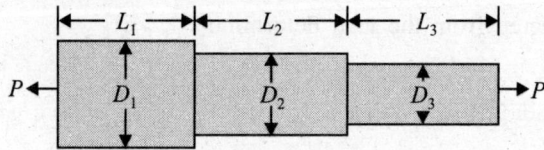

**Figure 12.9** Visual of Example 12.9.

Also show that if $D_1 = D_2 = D_3 = D$, and $L_1 + L_2 + L_3 = L$, Then $\Delta L = \dfrac{4PL}{\pi ED^2}$

then
$$\Delta L = \frac{4PL}{\pi ED^2} \qquad \text{(ii)}$$

*Solution* The stresses in 3 lengths of the bar are:

$$\sigma_1 = \frac{P}{\dfrac{\pi}{4} D_1^2} = \frac{4P}{\pi D_1^2}, \qquad \sigma_2 = \frac{P}{\dfrac{\pi}{4} D_2^2} = \frac{4P}{\pi D_2^2}, \qquad \sigma_3 = \frac{P}{\dfrac{\pi}{4} D_3^2} = \frac{4P}{\pi D_3^2}$$

Similarly extensions/elongations of the 3 bars are:

$$\Delta L_1 = \frac{PL_1}{A_1 E} = \frac{\sigma_1}{E} L_1 = \frac{4PL_1}{\pi D_1^{\,2} E}, \qquad \Delta L_2 = \frac{PL_2}{A_2 E} = \frac{\sigma_2}{E} L_2 = \frac{4PL_2}{\pi D_2^{\,2}}, \qquad \Delta L_3 = \frac{PL_3}{A_3 E} = \frac{\sigma_2}{E} L_3 = \frac{4PL_3}{\pi D_3^{\,2}}$$

Now total extension is:
$$\Delta L = \Delta L_1 + \Delta L_2 + \Delta L_3$$

or
$$\Delta L = \frac{4PL_1}{\pi D_1^2 E} + \frac{4PL_2}{\pi D_2^2 E} + \frac{4PL_3}{\pi D_3^2 E}$$

$\therefore$
$$\Delta L = \frac{4P}{\pi E} \left( \frac{L_1}{D_1^2} + \frac{L_2}{D_2^2} + \frac{L_3}{D_3^2} \right) \qquad \text{Proved} \qquad \text{(i)}$$

Let
$$D_1 = D_2 = D_3 = D$$

Above equation becomes

$$\Delta L = \frac{4P}{\pi E} \left( \frac{L_1}{D_1^{\,2}} + \frac{L_2}{D_2^{\,2}} + \frac{L_3}{D_3^{\,2}} \right)$$

or
$$\Delta L = \frac{4P}{\pi ED^2} (L_1 + L_2 + L_3) = \frac{4PL}{\pi ED^2} \qquad \text{Proved} \qquad \text{(ii)}$$

**EXAMPLE 12.10** A uniformly tapering rod of length $L$ from diameter $D_1$ to $D_2$ is subjected to pull $P$ as shown in Figure 12.10. If $\Delta L$ is the total extension of the bar and $E$ is the Young's modulus, show that

$$\Delta L = \frac{4PL}{\pi ED_1 D_2} \qquad \text{(i)}$$

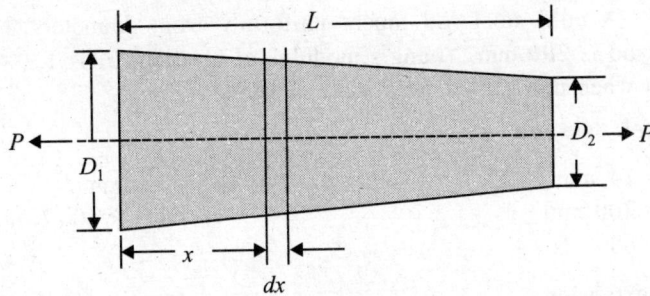

**Figure 12.10** Visual of Example 12.10.

**Solution** Consider an elemental strip of length $dx$ at a distance $x$ from end of bigger diameter $D_1$. Let the diameter at $x$ be $D'$.

Hence

$$D' = D_1 - \left(\frac{D_1 - D_2}{L}\right)x$$

Let

$$\left(\frac{D_1 - D_2}{L}\right) = C \qquad \therefore \ D' = D_1 - Cx$$

Cross-sectional area at $x = A' = \dfrac{\pi}{4}D'^2 = \dfrac{\pi}{4}(D_1 - Cx)^2$

Stress at section $x = \sigma' = \dfrac{P}{A'} = \dfrac{4P}{\pi(D_1 - Cx)^2}$

Strain at section $x = e' = \dfrac{\sigma'}{E} = \dfrac{4P}{\pi E(D_1 - Cx)^2}$

$\therefore$ Extension of the elemental strip $dx = e'dx = \dfrac{4P}{\pi E(D_1 - Cx)^2}dx$

$\therefore$ Total extension

$$\Delta L = \int_0^L \frac{4P}{\pi E(D_1 - Cx)^2}dx$$

or

$$\Delta L = \frac{4P}{\pi E}\int_0^L \frac{4P}{\pi E(D_1 - Cx)^2}dx$$

or

$$\Delta L = -\frac{4P}{\pi EC}\left[\frac{1}{D_1 - Cx}\right]_0^L = \frac{4P}{\pi EC}\left(\frac{1}{D_1 - CL} - \frac{1}{D_1}\right) \qquad \text{(i)}$$

But

$$C = \frac{D_1 - D_2}{L}$$

Substituting the value of $C$ in (i) and simplifying further,

$$\Delta L = \frac{4PL}{\pi ED_1D_2} \qquad \text{Proved.}$$

***EXAMPLE 12.11*** A mild steel rod tapers uniformly from diameters 24 mm to 12 mm. The length of the rod is 200 mm. Young's modulus of elasticity $E$ is $1.9 \times 10^5$ N/mm². Find the total extension when it is pulled by a force of 6000 N.

***Solution*** The given data are:

$$D_1 = 24 \text{ mm} \qquad\qquad D_2 = 12 \text{ mm}$$
$$L = 200 \text{ mm} \qquad\qquad E = 1.9 \times 10^5 \text{ N/mm}^2$$
$$P = 6000 \text{ N}$$

If $\Delta L$ is the total extension,
using Eq. (i)

$$\Delta L = \frac{4PL}{\pi E D_1 D_2} = \frac{4 \times 6000 \times 200}{\pi \times 1.9 \times 10^5 \times 24 \times 12}$$

$$\Delta L = 0.0279 \text{ mm} \qquad \text{(Answer)}$$

***EXAMPLE 12.12*** A compound member consists of steel tube of 160 mm internal diameter and 170 mm external diameter and an outer brass tube internal diameter 170 mm. The compound member of length 250 mm and subjected to an axial compressive load of 1000 kN. Find the outer diameter of the brass tube. $E_S = 2.1 \times 10^5$ N/mm², $E_B = 1.0 \times 10^5$ N/mm².

***Solution*** Area of steel tube $A_S = \dfrac{\pi}{4}(170^2 - 160^2) = 2591.814 \text{ mm}^2$

If $d$ is the outer diameter of the brass tube,

Area of brass tube $A_B = \dfrac{\pi}{4}(d^2 - 170^2)$

Axial load carried by the composite tube $P = 1000$ kN $= 1000000$ N
Length of each tube $L = 250$ mm
Young's modulus of steel $E_S = 2.1 \times 10^5$ N/mm²,
Young's modulus of brass $E_B = 1.0 \times 10^5$ N/mm²
Now Decrease in length of the compound tube = Decrease in length of brass tube = Decrease in length of steel tube

or
$$\Delta L = \Delta L_B = \Delta L_S$$

or
$$\Delta L_B = \Delta L_S$$

or
$$\frac{PL}{A_B E_B} = \frac{PL}{A_S E_S}$$

∴
$$A_B E_B = A_S E_S$$

or
$$A_B = \frac{E_S}{E_B} A = \frac{2.1 \times 10^5}{1 \times 10^5} \times 2591.814 = 5442.8094 \text{ mm}^2$$

or
$$\frac{\pi}{4}(d^2 - 170^2) = 5442.8094$$

Solving for external diameter of brass, $d = 189.288$ mm  (Answer)

## 12.8    RELATIONSHIP AMONG ELASTIC CONSTANTS

### 12.8.1    Relationship among $E$, $\mu$ and $G$

To establish the relationship among Young's modulus $E$, Poisson's ratio $\mu$ and modulus of rigidity, consider a square element of side $a$ subject to pure shear $\tau$ as shown in Figure 12.11. Due to the shear stress shown, let the element deform to $ADC'B'$. Draw perpendicular $BE$ to $B'D$. The shear strain shown in the diagram is $\phi$.

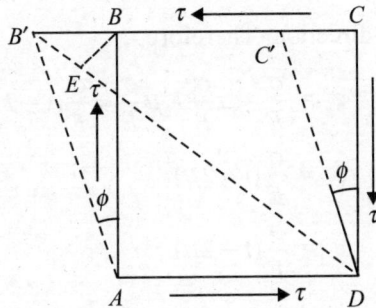

**Figure 12.11**    Element subject to shear.

Strain in diagonal $BD = \dfrac{DB' - DE}{DE} = \dfrac{B'E}{BD}$, since for a very small angle $\phi$, $ED = BD$.

But $BD = \sqrt{a^2 + a^2} = a\sqrt{2}$

Hence strain in diagonal $BD = \dfrac{B'E}{a\sqrt{2}}$                     (12.20)

Deformation is very small and hence, $B'E = B'B \cos 45° = a \tan\phi \cos 45°$
Substituting the value of $B'E$ using Eq. (12.20),

Strain in the diagonal $BD = \dfrac{a \tan\phi \cos 45°}{a\sqrt{2}} = \dfrac{1}{2}\tan\phi = \dfrac{1}{2}\phi$ since $\phi$ is very small

But from Eq. (12.17), $\phi = \dfrac{\tau}{G}$

Strain in the diagonal $BD = \dfrac{1}{2} \cdot \dfrac{\tau}{G}$ using Eq. (12.20)

Pure shear $\tau$ gives rise to tensile stress in the direction of $BD$ and compressive stress at right angles to it. These two normal stresses cause tension in diagonal $BD$.

Hence tensile strain in $BD = \dfrac{\tau}{E} + \mu\dfrac{\tau}{E} = \dfrac{\tau}{E}(1 + \mu)$                     (12.21)

Equating (12.20) and (12.21),

$$\dfrac{1}{2}\cdot\dfrac{\tau}{G} = \dfrac{\tau}{E}(1 + \mu)$$

$$E = 2G(1 + \mu) \qquad (12.22)$$

Equation (12.22) gives the relationship among $E$, $\mu$ and $G$.

### 12.8.2   Relationship among $E$, $\mu$ and $K$

A cubic element of size $(a \times a \times a)$, subject to identical stress in three mutually perpendicular directions is shown in Figure 12.12. The stress $\sigma$ in x-direction produces a tensile strain equal to $\dfrac{\sigma}{E}$ in the same direction, whereas the stresses in $y$ and $z$ directions produce compressive strain equal to $\mu\dfrac{\sigma}{E}$ each in x-direction. Therefore,

$$e_x = \frac{\sigma}{E} - \mu\frac{\sigma}{E} - \mu\frac{\sigma}{E} = \frac{\sigma}{E}(1 - 2\mu)$$

Similarly

$$e_y = \frac{\sigma}{E}(1 - 2\mu)$$

and

$$e_z = \frac{\sigma}{E}(1 - 2\mu)$$

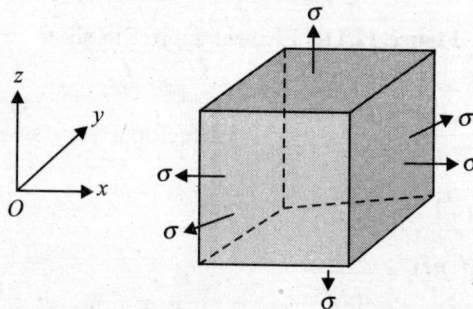

**Figure 12.12**   A cubic element subject to stress $\sigma$.

The summation of the above three strains gives the volumetric strain $e_v$. Hence

$$e_v = e_x + e_y + e_z = \frac{3\sigma}{E}(1 - 2\mu)$$

Definition of bulk modulus $K$ gives

$$K = \frac{\sigma}{e_v} = \frac{\sigma E}{3\sigma(1 - 2\mu)} = \frac{E}{3(1 - 2\mu)}$$

or

$$E = 3K(1 - 2\mu) \qquad (12.23)$$

Equation (12.23) gives the relationship of $E$, $K$ and $\mu$.

### 12.8.3   Relationship among $E$, $G$ and $K$

Eliminating $\mu$ from Eqs. (12.22) and (12.23), a relationship among $E$, $G$ and $K$ can be established.

From Eq. (12.22),

$$\mu = \frac{E}{2G} - 1$$

Substituting this value of $\mu$ in Eq. (12.23),

$$E = 3K\left[1 - 2\left(\frac{E}{2G} - 1\right)\right] = 3K\left(1 - \frac{E}{G} + 2\right) = 3K\left(3 - \frac{E}{G}\right) = 9K - \frac{3KE}{G}$$

or

$$E + \frac{3KE}{G} = 9K$$

or

$$E\left(1 + \frac{3K}{G}\right) = 9K$$

or

$$E = \frac{9KG}{G + 3K}$$

or

$$\frac{E}{9} = \frac{KG}{G + 3K}$$

or

$$\frac{9}{E} = \frac{G + 3K}{KG}$$

$\therefore$

$$\frac{9}{E} = \frac{1}{K} + \frac{3}{G} \qquad (12.24)$$

equation gives the relationship among $E$, $G$ and $K$.

**EXAMPLE 12.13** Determine the Poisson's ratio ($\mu$) and bulk modulus ($K$) if the bar has a modulus of elasticity ($E$) equal to $1.98 \times 10^5$ N/mm$^2$ and modulus of rigidity ($G$) is $0.85 \times 10^5$ N/mm$^2$.

**Solution** Given data are:

$$E = 1.98 \times 10^5 \text{ N/mm}^2$$
$$G = 0.85 \times 10^5 \text{ N/mm}^2$$

We have now Eq. (12.23) as:

$$E = 3K(1 - 2\mu)$$

And Eq. (12.24) respectively as:

$$\frac{9}{E} = \frac{1}{K} + \frac{3}{G}$$

Substituting the values of $K$ and $G$ in Eq. (12.24)

$$\frac{9}{1.98 \times 10^5} = \frac{1}{K} + \frac{3}{0.85 \times 10^5}$$

or

$$0.00004545 = \frac{1}{K} + 0.00003529$$

or
$$\frac{1}{K} = 0.00001026$$

$\therefore$                    $K = 0.97456 \times 10^5$ N/mm$^2$    (Answer)

Substituting the values of $K$ and $E$ in Eq. (12.23),

$$1.98 \times 10^5 = 3 \times 0.97465 \times 10^5 (1 - \mu)$$

or                $(1 - \mu) = \dfrac{1.98 \times 10^5}{3 \times 0.97465 \times 10^5} = 0.6771$

$\therefore$                    $m = 0.3229 = \dfrac{1}{3.097}$, i.e. $m = 3.097$    (Answer)

**EXAMPLE 12.14**   The modulus of rigidity $G$ of a steel bar is $4.8 \times 10^4$ N/mm$^2$ and the Poisson's ratio is 0.25. Determine the Young's modulus $E$ and bulk modulus $K$.

**Solution**   Using Eq. (12.22),

$$E = 2G (1 + \mu)$$
$$E = 2 \times 4.8 \times 10^4 (1 + 0.25)$$
$\therefore$                    $E = 1.2 \times 10^5$ N/mm$^2$    (Answer)

To obtain the value of $K$, use Eq. (12.24), i.e.

$$\frac{9}{E} = \frac{1}{K} + \frac{3}{G}$$

Substituting the values of $E$ and $G$,

$$\frac{9}{1.2 \times 10^5} = \frac{1}{K} + \frac{3}{4.8 \times 10^4}$$

or                $0.000075 = \dfrac{1}{K} + 0.0000625$

or                    $\dfrac{1}{K} = 0.0000125$

$\therefore$                    $K = 0.8 \times 10^5$ N/mm$^2$    (Answer)

**EXAMPLE 12.15**   A bar 30 mm in diameter is subjected to a tensile load of 60 kN and the measured extension on 300 mm gauge length is 0.15 mm and change in diameter is 0.003 mm. Calculate Poisson's ratio. If $E = 2 \times 10^5$ N/mm$^2$, find also bulk bodulus of the material.

**Solution**   Poisson's ratio is given by

$$\mu = \frac{\text{Lateral strain}}{\text{Linear strain}} = \frac{e_y}{e_x}$$

Now from the given data,

$$e_x = \frac{\Delta L}{L} = \frac{0.15}{300} = 5 \times 10^{-4}$$

and
$$e_y = \frac{\Delta d}{d} = \frac{0.003}{30} = 1 \times 10^{-4}$$

Hence Poisson's ratio is:

$$\mu = \frac{e_y}{e_x} = \frac{1 \times 10^{-4}}{5 \times 10^{-4}} = \frac{1}{5} = 0.25 \quad \text{(Answer)}$$

The relationship among $E$, $K$ and $\mu$ is given by Eq. (12.23), i.e.

$$E = 3K(1 - 2\mu)$$

Substituting the values of $E$ and $\mu$

$$2 \times 10^5 = 3K (1 - 2 \times 0.25)$$

$$\therefore \quad K = \frac{2 \times 10^5}{3 \times 0.50} = 1.33333 \times 1.0^5 \text{ N/mm}^2$$

**EXAMPLE 12.16** A bar of cross-section 8 mm × 8 mm is subjected to an axial pull of 7 kN. The lateral dimension of the bar is found to be changed to 7.9985 mm × 7.9985 mm. If the modulus of rigidity of the material is $0.8 \times 10^5$ N/mm², determine the Poisson's ratio and modulus of elasticity.

**Solution** From the given data,

$$e_y = \frac{8 - 7.9985}{8} = 1.875 \times 10^{-4}$$

We know Poisson's ratio

$$\mu = \frac{e_y}{e_x}$$

But longitudinal strain $e_x$ cannot be obtained as the longitudinal length and extension are not given. Equation (12.14) gives

$$\sigma = Ee_x$$

or
$$e_x = \frac{\sigma}{E} = \frac{\dfrac{P}{A}}{E} = \frac{\dfrac{7000}{8 \times 8}}{E} = \frac{109.375}{E}$$

But
$$e_y = \mu e_x = \mu \frac{109.375}{E}$$

or
$$1.875 \times 10^{-4} = 109.375 \frac{\mu}{E}$$

or
$$E = 583333.333\mu \quad \text{(i)}$$

Using Eq. (12.22),
$$E = 2G (1 + \mu) = 2 \times 0.8 \times 10^5(1 + \mu)$$

or
$$583333.333\mu = 160000(1 + \mu)$$

or
$$(583333.333 - 160000)\mu = 160000$$

$$\therefore \qquad \mu = \frac{160000}{423333.333} = 0.3779 \qquad \text{(Answer)}$$

Substituting the value of $\mu$ in (i)

$$E = 583333.333 \times 0.3779 = 2.2044 \times 10^5 \text{ N/mm}^2 \qquad \text{(Answer)}$$

## 12.9   CONCLUSION

Analysis of compressive, tensile and shear stresses are presented in the beginning of the chapter with figures that are easy to understand. Then linear compressive, tensile, shear and volumetric strains with equations are defined. Equations of volumetric strain for cylindrical, rectangular, spherical bar are derived. Elastic limit of material, stress-strain relationship, Hooke's Law are explained. All the elastic constants are defined and the relationships among these constants and with Poisson's ratio are derived. A number of worked-out examples are presented to clarify the theory.

## EXERCISES

**12.1** How do you define strength of materials? Explain compressive, tensile and shear stress by simple diagram.

**12.2** Describe tensile, compressive, shear and volumetric strain with simple sketches where necessary.

**12.3** Derive expressions for volumetric strain of cylinder and sphere.

**12.4** Define stress and strain, and state Hooke's Law. Drawing a neat sketch of stress-strain diagram, explain the elastic behaviour of mild steel under tension.

**12.5** Name the different elastic constants. Give the relationship among the elastic constants.

**12.6** A steel bar of 20 mm diameter and 2.5 m long is subjected to a pull of 100 kN. Calculate the elongation of the bar in mm. Take $E = 2 \times 10^5$ N/mm$^2$.

$$\text{(Answer   3.978 mm)}$$

**12.7** Find the Young's modulus of a rod of diameter 30 mm and of length 400 mm which is subjected to an axial tensile load of 100 kN and the extension of the rod due to force is 0.8 mm.   (Answer   $0.70735 \times 10^5$ N/mm$^2$)

**12.8** A bar 4 m long and $(100 \times 200)$ mm$^2$ in sections is subjected to a pull of 50 kN. If Young's modulus of the material is $2 \times 10^5$ N/mm$^2$, find
  (i) Stress set up in the bar material
  (ii) Strain
  (iii) Elongation of the bar

$$\text{(Answer   (i) Stress} = 2.5 \text{ N/mm}^2\text{, (ii) Strain} = 1.25 \times 10^{-5},$$
$$\text{(iii) Elongation} = 0.05 \text{ mm)}$$

# CHAPTER 13

# Generalized Stress and Strain

## 13.1  INTRODUCTION

In Chapter 12, discussion on stresses and strains was confined when the force acts on a plane which is at right angles to the direction of application of the force. In many engineering problems, compressive, tensile and shear stresses are acting at the same time. In such situations, resultant stress across on any plane is neither normal nor tangential to the plane in which they act. In this chapter, analysis of stresses will be made when the applied force acts on an inclined or oblique section.

## 13.2  STRESSES ON INCLINED SECTION

### 13.2.1  Member Subjected to a Direct Stress

Consider the rectangular member $ABCD$ of uniform cross-sectional area $A$ and of unit thickness in Figure 13.1.

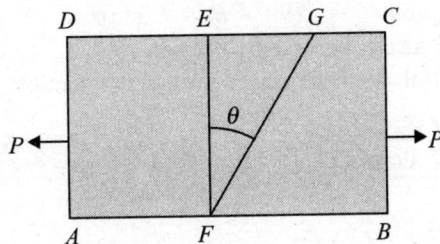

**Figure 13.1**  A rectangular bar $ABCD$ unit width.

Consider the section $EF$ on the member which is perpendicular to the line of action of the applied force $P$.

Then stress on the section $EF$ is given as:

$$\sigma = \frac{\text{Applied force}}{\text{Area}} = \frac{P}{EF \times 1} = \frac{P}{A} \tag{13.1}$$

Now consider another section $FG$ at an angle $\theta$ to $EF$ as shown in Figure 13.1.

$$\text{Stress on this section} = \frac{\text{Force}}{\text{Area of section } FG}$$

Area of section
$$FG = \frac{EF}{\cos\theta} = \frac{A}{\cos\theta}$$

$$\because \quad \frac{EF}{FG} = \cos\theta \qquad\qquad \therefore \ FG = \frac{EF}{\cos\theta}$$

Stress on the inclined section $FG = \dfrac{P}{\left(\dfrac{A}{\cos\theta}\right)} = \dfrac{P}{A}\cos\theta = \sigma\cos\theta \tag{13.2}$

This stress $\sigma\cos\theta$ or force $P$ may be resolved into two components on the section $FG$. They are $P_n$ and $P_t$ which are normal and tangential to $FG$ as shown in Figure 13.2.

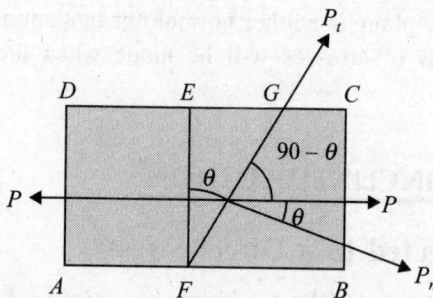

**Figure 13.2** Component of $P$ or $\sigma\cos\theta$ on $FG$.

Now $\qquad P_n = P\cos\theta$ and $P_t = P(90 - \theta) = P\sin\theta$

Let $\qquad \sigma_n = $ Normal stress across the section $FG$

and $\qquad \sigma_t = $ Tangential or shear stress across the section $FG$

Now $\qquad \sigma_n = \dfrac{\text{Force } P_n}{\text{Area of along } FG} = \dfrac{P\cos\theta}{\left(\dfrac{A}{\cos\theta}\right)} = \dfrac{P}{A}\cos^2\theta = \sigma\cos^2\theta \quad \because \dfrac{P}{A} = \sigma$

$$\therefore \qquad \sigma_n = \sigma\cos^2\theta \tag{13.3}$$

and $\qquad \sigma_n = \dfrac{\text{Force } P_t}{\text{Area on } FG} = \dfrac{P_t}{\left(\dfrac{A}{\cos\theta}\right)} = \dfrac{P\sin\theta}{\left(\dfrac{A}{\cos\theta}\right)} = \dfrac{P}{A}\sin\theta\cos\theta = \sigma\sin\theta\cos\theta$

or $\qquad \sigma_t = \sigma\sin\theta\cos\theta = \dfrac{\sigma}{2} \times 2\sin\theta\cos\theta = \dfrac{\sigma}{2}\sin 2\theta \tag{13.4}$

It can be observed that in Eq. (13.3), $\sigma_n$ will be maximum when $\cos^2\theta$ or $\cos\theta$ is maximum, i.e. when $\theta = 0°$.

Thus maximum normal stress is:

$$\sigma_n = \sigma \cos^2 0° = \sigma \qquad (13.5)$$

It has also been observed from Eq. (13.4) that shear stress $\sigma_t$ across $FG$ is the maximum when $\sin 2\theta$ maximum, i.e. is $\sin 2\theta = 1$

or $\qquad\qquad\qquad 2\theta = 90° \quad$ or $\quad 270°$

$\therefore \qquad\qquad\qquad \theta = 45° \quad$ or $\quad 135°$

It means that shear or tangential stress is the maximum when the plane $FG$ is inclined at 45° and 135° to plane $EF$.

$\therefore$ Maximum value of shear or tangential stress $= \sigma_t = \dfrac{\sigma}{2}\sin 90° = \dfrac{\sigma}{2} \qquad (13.6)$

## 13.2.2  Member Subjected to Two Direct Tensile Stresses

A block $ABCD$ of unit thickness is considered. This is subjected to two principal tensile stresses $\sigma_1$ and $\sigma_2$ as shown in Figure 13.3.

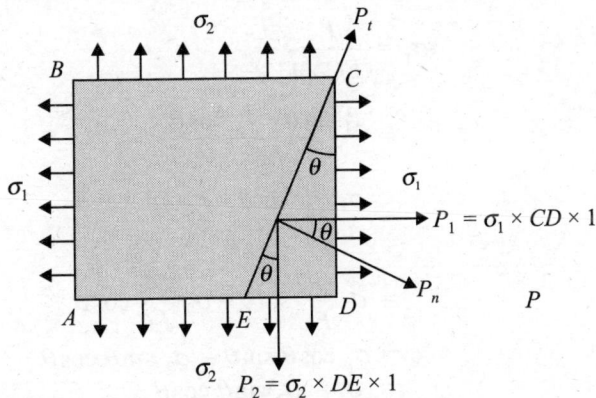

**Figure 13.3**  Member subjected to two direct tensile stresses.

Consider the oblique plane $CE$ at an angle of $\theta$ with principal plane $CD$.
Now total force normal to the plane $CE$ is:

$$P_n = P_1 \cos\theta + P_2 \sin\theta$$

And total force tangential to the plane $CE$ is:

$$P_t = P_1 \sin\theta + \cos\theta$$

Normal stress on plane $CE$:

$$\sigma_n = \frac{P_n}{CE \times 1}$$

$$= \frac{P_1 \cos\theta + P_2 \sin\theta}{CE}$$

$$= \frac{\sigma_1 \, CD \cos\theta + \sigma_2 \, DE \sin\theta}{CE}$$

$$= \sigma_1 \frac{CD}{CE} \cos\theta + \sigma_2 \frac{DE}{CE} \sin\theta$$

$$= \sigma_1 \cos\theta \cos\theta + \sigma_2 \sin\theta \sin\theta$$

$$= \sigma_1 \cos^2\theta + \sigma_2 \sin^2\theta$$

$$= \sigma_1 \left( \frac{1 + \cos 2\theta}{2} \right) + \sigma_2 \left( \frac{1 - \cos 2\theta}{2} \right)$$

$$\because \qquad \cos^2\theta = \frac{1 + \cos 2\theta}{2} \text{ and } \sin^2\theta = \frac{1 - \cos 2\theta}{2}$$

$$\therefore \qquad \sigma_n = \frac{\sigma_1 + \sigma_2}{2} + \frac{\sigma_1 - \sigma_2}{2} \cos 2\theta \qquad\qquad (13.7)$$

Tangential or shear stress along $EC$:

$$\sigma_1 = \frac{P_t}{EC \times 1}$$

$$= \frac{P_1 \sin\theta - P_2 \cos\theta}{EC}$$

$$= \frac{\sigma_1 \, CD \sin\theta - \sigma_2 \, DE \cos\theta}{EC}$$

$$= \sigma_1 \frac{CD}{EC} \sin\theta - \sigma_2 \frac{DE}{EC} \cos\theta$$

$$= \sigma_1 \cos\theta \sin\theta - \sigma_2 \sin\theta \cos\theta$$

$$= (\sigma_1 - \sigma_2)\sin\theta \cos\theta$$

$$= (\sigma_1 - \sigma_2) \frac{2 \sin\theta \cos\theta}{2}$$

$$\therefore \qquad \sigma_t = \left( \frac{\sigma_1 - \sigma_2}{2} \right) \sin 2\theta \qquad\qquad (13.8)$$

Similarly, it can be shown that if the stress $\sigma_1$ is tensile and $\sigma_2$ is compressive (both unlike), Eqs. (13.7) and (13.8) become:

$$\sigma_n = \frac{\sigma_1 - \sigma_2}{2} + \frac{\sigma_1 + \sigma_2}{2} \cos 2\theta \qquad\qquad \text{(using Eq. (13.7))}$$

and

$$\sigma_t = \left( \frac{\sigma_1 + \sigma_2}{2} \right) \sin 2\theta \qquad\qquad \text{(using Eq. (13.8))}$$

Resultant stress on the section $CE$:

$$\sigma_R = \sqrt{\sigma_n^2 + \sigma_t^2} \qquad (13.9)$$

The angle that the line of action of resultant stress makes with the plane is called *obliquity*. If the angle is $\phi$ then:

$$\tan \phi = \frac{\sigma_t}{\sigma_n} \qquad (13.10)$$

Shear stress given by equation is maximum when $\sin 2\theta$ in Eq. (13.8) is equal to 1.
    Maximum shear stress is:

$$\sigma_{t\,max} = \frac{\sigma_1 - \sigma_2}{2} \qquad (13.11)$$

Principal planes are the planes where shear stress is zero. Equating Eq. (13.8) of shear to zero:

$$\therefore \qquad \sigma_t = \left( \frac{\sigma_1 - \sigma_2}{2} \right) \sin 2\theta = 0$$

Since $(\sigma_1 - \sigma_2)$ cannot be zero, $\therefore$ $\sin 2\theta = 0$

or
$$2\theta = 0° \quad \text{or} \quad 180°$$
or
$$\theta = 0° \quad \text{or} \quad 90°$$

When $\theta = 0°$, Eq. (13.7) becomes:

$$\sigma_n = \sigma_1 \qquad (13.12)$$

When $\theta = 90°$, Eq. (13.7) becomes

$$\sigma_n = \sigma_2 \qquad (13.13)$$

**EXAMPLE 13.1**   A circular bar is subjected to an axial pull of $P = 150$ kN. The maximum allowable shear stress on any section is 60 N/mm$^2$. Find the diameter of the bar.

**Solution**   Let $d$ be the required diameter of the bar
Hence area of the bar

$$A = \frac{\pi}{4} d^2$$

Direct stress by Eq. (13.1) is

$$\sigma = \frac{P}{A} = \frac{150 \times 1000}{\frac{\pi}{4} d^2} = \frac{600000}{\pi d^2} \text{ N/mm}^2$$

Maximum shear stress is given by Eq. (13.6) $= \dfrac{\sigma}{2} = 60 \text{ N/mm}^2 = \dfrac{600000}{2 \times \pi d^2}$

or
$$60 \times 2 \times \pi d^2 = 600000$$

or
$$d^2 = \frac{600000}{2 \times \pi \times 60} = 1591.5494$$

$\therefore$
$$d = 39.894 \text{ mm} \qquad \text{(Answer)}$$

**EXAMPLE 13.2**   An axial pull of 22 kN is applied to a rectangular bar of cross-sectional area 11500 mm². Determine the normal and shear stresses on a section which is inclined at an angle of 30° with the normal cross section of the bar.

**Solution**   Data given are:

$$P = 22 \text{ kN} = 22000 \text{ N} \quad A = 11500 \text{ mm}^2, \quad \theta = 30°$$

Using Eq. (13.1)

$$\sigma = \frac{22000}{11500} = 1.913 \text{ N/mm}^2$$

Using Eq. (13.3) for normal stress:

$$\sigma_n = \sigma \cos^2 \theta = \sigma \cos^2 30° = 1.913 \times 0.75 = 1.43475 \text{ N/mm}^2 \qquad \text{(Answer)}$$

Similarly, using Eq. (13.4) for shear stress:

$$\sigma_t = \frac{\sigma}{2} \sin 2\theta = \frac{1.913}{2} \sin 60° = 0.9565 \times 0.866 = 0.8283 \text{ N/mm}^2 \qquad \text{(Answer)}$$

**EXAMPLE 13.3**   A rectangular bar of sectional area 1000 m² is subjected to a tensile force $P$ as shown in Figure 13.4. The permissible normal and shear stresses on the oblique plane $EF$ are given as 8 N/mm² and 4 N/mm² respectively. Determine the safe value of $P$.

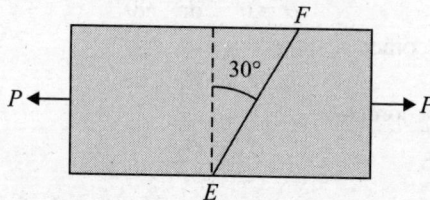

**Figure 13.4**   Visual of Example 13.3.

**Solution**   Data given in the problem are:

$$A = 1000 \text{ m}^2 \qquad \sigma_n = 8 \text{ N/mm}^2 \qquad \sigma_t = 4 \text{ N/mm}^2 \qquad \theta = 30°$$

Let $\sigma$ be the safe stress in the member
Using Eq. (13.3) for normal stress:

$$\sigma_n = \sigma \cos^2\theta = \sigma \cos^2 30°$$
$$8 = \sigma \cos^2 30° = \sigma \times 0.75$$

∴

$$\sigma = \frac{8}{0.75} = 10.667 \text{ N/mm}^2$$

Again using Eq. (13.4)

$$\sigma_t = \frac{\sigma}{2} \sin 2\theta$$

$$4 = \frac{\sigma}{2} \sin 60° = \frac{\sigma}{2} \times 0.866$$

$\therefore$ $$\sigma = 9.237 \text{ N/mm}^2$$

Now least of the above two is considered to be safe, i.e. $\sigma = 9.237 \text{ N/mm}^2$
Hence safe axial force $P = 9.237 \times 1000 = 9237 \text{ N}$    (Answer)

**EXAMPLE 13.4**  Two wooden pieces 12 cm × 12 cm in cross section are glued together along the line *EF* as shown in Figure 13.5. If the allowable shearing stress along *EF* is 1.0 N/mm², what will be the value of maximum pull *P*?

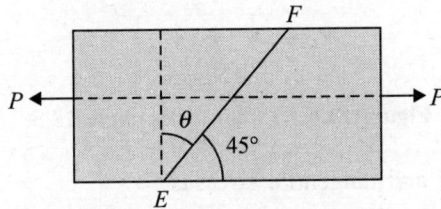

**Figure 13.5**  Visual of Example 13.4.

**Solution**  Given data
Cross-sectional area

$$A = 12 \times 12 = 144 \text{ cm}^2 = 14400 \text{ mm}^2$$

Allowable shear stress along *EF*

$$\sigma_1 = 1.0 \text{ N/mm}^2$$

And from Figure

$$\theta = 45°$$

Let $\sigma$ be the maximum allowable stress in the direction of pull *P*
Using Eq. (13.4),

$$\sigma_t = \frac{\sigma}{2} \sin 2\theta$$

or

$$1 = \frac{\sigma}{2} \sin 90° = \frac{\sigma}{2} \times 1$$

$\therefore$ $$\sigma = 2 \text{ N/mm}^2$$

Hence maximum pull $P = \sigma \times A = 2 \times 14400 = 28800 \text{ N} = 28.8 \text{ kN}$    (Answer)

**EXAMPLE 13.5**  The stresses at a point in a bar are 180 N/mm² (tensile) and 80 N/mm² (compressive). Calculate the resultant stress in magnitude and direction on a plane inclined at 60° to the axis of the major stress. Also determine the maximum intensity of shear stress in the material at the point. Find further the intensity of stress which acting alone can produce the same maximum strain if Poisson's ratio is 0.25.

**Solution**  Visual of Example 13.5 is shown in Figure 13.6.
Data given are:
Major principal stress = $\sigma_1$ = 180 N/mm²
Minor principal stress = $\sigma_2$ = −80 N/mm² (minus as compressive)
Angle $\theta = (90° - 60°) = 30°$

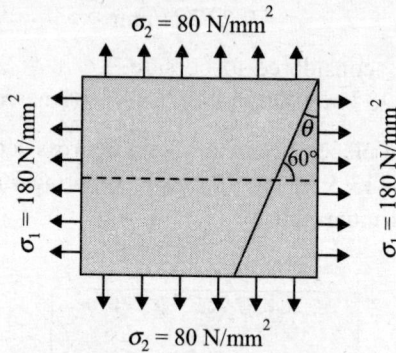

**Figure 13.6** Visual of Example 13.5.

Now evaluate the normal and tangential stresses.
Using Eq. (13.7),

$$\sigma_n = \frac{180 + (-80)}{2} + \frac{180 - (-80)}{2} \cos 60°$$
$$= 50 + 140 \times 0.5$$
$$= 120 \ \text{N/mm}^2$$

Using Eq. (13.8),

$$\sigma_t = \left( \frac{\sigma_1 - \sigma_2}{2} \right) \sin 2\theta$$

$$= \frac{180 - (-80)}{2} \sin 60°$$
$$= 140 \times 0.866025$$
$$= 121.243 \ \text{N/mm}^2$$

Use Eq. (13.9) to evaluate the resultant stress,

$$\sigma_R = \sqrt{\sigma_n^2 + \sigma_t^2}$$

$$= \sqrt{120^2 + 121.243^2}$$
$$= \sqrt{29100}$$
$$= 170.587 \ \text{N/mm}^2 \qquad \text{(Answer)}$$

Direction of the resultant stress is given by Eq. (13.10).

$\tan\phi = \dfrac{\sigma_t}{\sigma_n}$, where $\phi$ is the angle that the line of action of resultant stress makes with the plane.

$$\tan\phi = \frac{121.243}{120} = 1.0103$$
$$\phi = \tan^{-1} 1.01103 = 45.295° \qquad \text{(Answer)}$$

Maximum shear stress is given by equation

$$\sigma_{t\,max} = \frac{\sigma_1 - \sigma_2}{2}$$

$$= \frac{180 - (-80)}{2}$$

$$= 130 \text{ N/mm}^2 \quad \text{(Answer)}$$

**EXAMPLE 13.6**  The principal tensile stresses at a point in a stained bar across two perpendicular planes are 100 N/mm² (tensile) and 50 N/mm². Calculate the normal stress, shear stress and resultant stress on a plane inclined at 30° to the axis of the major stress. Determine also the obliquity. Find further the intensity of stress which acting alone can produce the same maximum strain if Poisson's ratio is 0.25.

**Solution**  Data given are:

$$\text{Major principal stress} = \sigma_1 = 100 \text{ N/mm}^2$$
$$\text{Minor principal stress} = \sigma_2 = 50 \text{ N/mm}^2$$
$$\text{Angle } \theta = 30°$$
$$\text{Poisson's ratio } \mu = 0.3$$

Now evaluate the normal, shear and resultant stresses.
Using Eq. (13.7),

$$\sigma_n = \frac{\sigma_1 + \sigma_2}{2} + \frac{\sigma_1 - \sigma_2}{2} \cos 2\theta$$

$$\sigma_n = \frac{100 + 50}{2} + \frac{100 - 50}{2} \cos 60°$$

$$= 75 + 25 \times 0.5$$

$$= 87.5 \text{ N/mm}^2 \quad \text{(Answer)}$$

Using Eq. (13.8),

$$\sigma_t = \left(\frac{\sigma_1 - \sigma_2}{2}\right) \sin 2\theta$$

$$= \frac{100 - 50}{2} \sin 60°$$

$$= 25 \times 0.866025$$

$$= 21.65 \text{ N/mm}^2 \quad \text{(Answer)}$$

Use Eq. (13.9) to evaluate the resultant stress.

$$\sigma_R = \sqrt{\sigma_n^2 + \sigma_t^2}$$

$$= \sqrt{87.5^2 + 21.65^2}$$

$$= \sqrt{8124.9725}$$

$$= 90.1386 \text{ N/mm}^2 \quad \text{(Answer)}$$

Obliquity $\phi$ is given by Eq. (13.10).

$$\tan\phi = \frac{\sigma_t}{\sigma_n}$$

$$\tan\phi = \frac{21.65}{87.5} = 0.247428$$

$$\phi = \tan^{-1} 0.247428 = 13.897 \qquad \text{(Answer)}$$

Let $\sigma$ be the stress which acting alone will produce the same maximum strain. The maximum strain will be in the direction of major principal stress.

$$\text{Maximum strain} = \frac{\sigma_1}{E} - \frac{\mu\sigma_2}{E} = \frac{1}{E}(\sigma_1 - \mu\sigma_2)$$

$$= \frac{1}{E}(100 - 0.3 \times 50)$$

$$= \frac{85}{E} \qquad\qquad\text{(i)}$$

$$\text{Strain due to stress} = \frac{\sigma}{E} \qquad\qquad\text{(ii)}$$

Equating (i) and (ii),

$$\frac{85}{E} = \frac{\sigma}{E}$$

$$\therefore \qquad\qquad \sigma = 85 \text{ N/mm}^2 \qquad \text{(Answer)}$$

**EXAMPLE 13.7**   A block of 6 cm × 4 cm × 2 cm is subjected to uniformly distributed tensile forces of resultants 1500 N and 720 N as shown in Figure 13.7. Compute the normal and shear stresses developed along the diagonal $DB$.

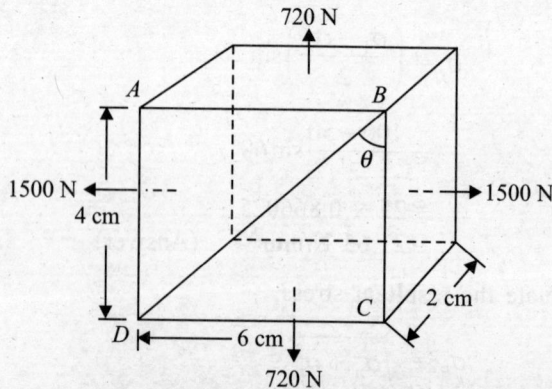

**Figure 13.7**   Visual of Example 13.7.

**Solution**   The length, breadth, depth and applied forces in $x$ and $y$ directions are shown in Figure 13.7.

Cross-sectional area of the block in $x$-direction = $4 \times 2 = 8$ cm$^2$
Cross-sectional area of the block in $y$-direction = $6 \times 2 = 12$ cm$^2$

$$\text{Stress along } x\text{-axis} = \frac{\text{Force in } x\text{-axis}}{\text{Area normal to } x\text{-axis}}$$

$$= \frac{1500}{8} = 1875 \text{ N/cm}^2$$

$$\therefore \quad \sigma_1 = 1875 \text{ N/cm}^2$$

$$\text{Stress along } y\text{-axis} = \frac{\text{Force in } y\text{-axis}}{\text{Area normal to } y\text{-axis}}$$

$$= \frac{720}{12} = 60 \text{ N/cm}^2$$

$$\therefore \quad \sigma_2 = 60 \text{ N/cm}^2$$

From Figure (13.7),

$$\tan \theta = \frac{6}{4} = 1.5$$

$$\theta = \tan^{-1} 1.5 = 56.31°$$

Let $\sigma_n$ and $\sigma_t$ be the normal and shear stresses on diagonal $BD$.
Using Eq. (13.7),

$$\sigma_n = \frac{\sigma_1 + \sigma_2}{2} + \frac{\sigma_1 - \sigma_2}{2} \cos 2\theta$$

$$= \frac{1875 + 60}{2} + \frac{1875 - 60}{2} \cos (2 \times 56.31)°$$

$$= 967.5 + 907.5 \times (-0.384617)$$

$$= 618.4595 \text{ N/cm}^2 \quad \text{(Answer)}$$

Using Eq. (13.8),

$$\sigma_t = \left( \frac{\sigma_1 - \sigma_2}{2} \right) \sin 2\theta$$

$$= \frac{(1875 - 60)}{2} \sin(2 \times 56.31)°$$

$$= 907.5 \times 0.923076$$

$$= 837.691 \text{ N/cm}^2 \quad \text{(Answer)}$$

## 13.2.3  Member Subjected to a Simple Shear Stress

Figure 13.8 shows a rectangular bar of uniform cross sectional area $A$ of unit thickness. The bar is subjected to simple shear stress $\tau$ across the faces $BC$ and $AD$. Let $BE$ the oblique section at angle $\theta$ on which normal and tensile are developed.

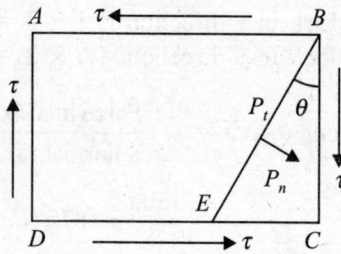

**Figure 13.8** Member subjected to simple shear stress $\tau$.

We know that shear stress is always accompanied by an equal shear stress at right angle to it as shown in Figure 13.8. These forces can be resolved along the inclined surface and normal to the inclined surface. Then considering forces acting on the wedge $EBC$, $P_t$ and $P_n$ are the tangential or shear force along $EB$ and $P_n$ is the normal forces to $EB$.

From this normal force $P_n$, it can be shown the relation between $\sigma_n$ and $\tau$ in similar process after simplification as:

$$\sigma_n = \tau \sin 2\theta \tag{13.14}$$

Similarly relationship between $\tau_t$ and $\tau$ can be obtained after simplification as:

$$\sigma_t = -\tau \cos 2\theta \tag{13.15}$$

### 13.2.4 Direct Stresses in Two Perpendicular Directions Accompanied by Simple Shear Stress

In Figure 13.9, the member in Figure 13.8 is further subjected to the two mutually perpendicular tensile stresses $\sigma_1$ and $\sigma_2$. In this situation, equation of normal stress $\sigma_n$ will be the combination of Eqs. (13.7) and (13.14). It can be simplified to obtain normal stress in this situation as:

$$\sigma_n = \frac{\sigma_1 + \sigma_2}{2} + \frac{\sigma_1 - \sigma_2}{2} \cos 2\theta + \tau \sin 2\theta \tag{13.16}$$

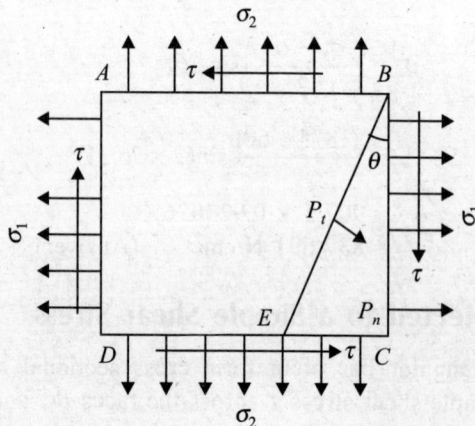

**Figure 13.9** Member subjected to two direct perpendicular stresses accopanied by simple shear stress $\tau$.

Similarly tensile stress $\sigma_t$ is combination of Eqs. (13.8) and (13.15). It can be simplified to obtain tensile stress in this situation as:

$$\sigma_t = \left(\frac{\sigma_1 - \sigma_2}{2}\right)\sin 2\theta - \tau\cos 2\theta \tag{13.17}$$

Substituting $\sigma_t = 0$ Eq. (13.17) in principal planes and further simplification on Eq. (13.16), maximum and minimum principal stresses are obtained as:

$$\text{Maximum principal stress} = \frac{\sigma_1 + \sigma_2}{2} + \sqrt{\left(\frac{\sigma_1 - \sigma_2}{2}\right)^2 + \tau^2} \tag{13.18}$$

$$\text{Minimum principal stress} = \frac{\sigma_1 + \sigma_2}{2} - \sqrt{\left(\frac{\sigma_1 - \sigma_2}{2}\right)^2 + \tau^2} \tag{13.19}$$

Maximum shear stress is obtained by differentiating Eq. (13.16) and equating to zero. This condition gives after simplification:

$$\tan 2\theta = \frac{2\tau}{\sigma_1 - \sigma_2} \tag{13.20}$$

Then

$$\sin 2\theta = \pm\frac{\sigma_2 - \sigma_1}{\sqrt{(\sigma_2 - \sigma_1)^2 + 4\tau^2}} \tag{13.21}$$

and

$$\cos 2\theta = \pm\frac{2\tau}{\sqrt{(\sigma_2 - \sigma_1)^2 + 4\tau^2}} \tag{13.22}$$

Substituting the values of $\sin 2\theta$ and $\cos 2\theta$ in Eq. (13.17), maximum shear stress $(\sigma_{t\,\text{max}})$ is obtained after simplification as:

$$\sigma_{t\,\text{max}} = \frac{1}{2}\sqrt{(\sigma_1 - \sigma_2)^2 + 4\tau^2} \tag{13.23}$$

***EXAMPLE 13.8***  Two mutually perpendicular tensile stresses of 120 N/mm$^2$ and 60 N/mm$^2$ are acting a point of a rectangular bar as shown in Figure 13.10. Each of the above stresses is accompanied by a shear stress of 90 N/mm$^2$. Determine the normal stress, shear stress and resultant stress on a plane *AB* inclined at 40° with axis of minor tensile stress as shown in the same figure.

***Solution***  Major and minor tensile stresses, shear stress and angle with oblique plane *AB* are clearly shown in Figure 13.10.

To determine the normal stress $\sigma_n$, use Eq. (13.16).

$$\sigma_n = \sigma_n = \frac{\sigma_1 + \sigma_2}{2} + \frac{\sigma_1 - \sigma_2}{2}\cos 2\theta + \tau\sin 2\theta$$

$$= \frac{120 + 60}{2} + \frac{120 - 60}{2}\cos 80° + 90\sin 80°$$

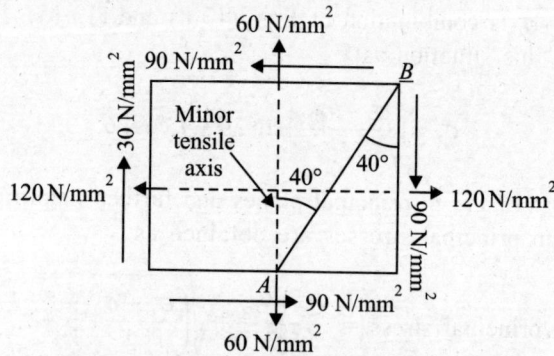

**Figure 13.10** Visual of Example 13.8.

$$= 90 + 30 \cos 80° + 90 \sin 80°$$
$$= 90 + 5.209 + 88.632$$
$$= 183.841 \text{ N/mm}^2 \quad \text{(Answer)}$$

To determine the shear or tangential stress $\sigma_t$, use Eq. (13.17).

$$\sigma_t = \left(\frac{\sigma_1 - \sigma_2}{2}\right) \sin 2\theta - \tau \cos 2\theta$$

$$= \frac{120 - 60}{2} \sin 80° - 90 \cos 80°$$

$$= 29.5442 - 15.6283$$

$$= 13.9159 \text{ N/mm}^2 \quad \text{(Answer)}$$

Resultant stress is obtained by using Eq. (13.9).

$$\sigma_R = \sqrt{\sigma_n^2 + \sigma_t^2}$$

$$= \sqrt{183.841^2 + 13.9159^2}$$

$$= \sqrt{33991.16555}$$

$$= 184367 \text{ N/mm}^2 \quad \text{(Answer)}$$

***EXAMPLE 13.9*** A rectangular block of material is subjected to two tensile stresses of magnitude 148 N/mm$^2$ and 60 N/mm$^2$ along two mutually perpendicular planes as shown in Figure 13.11. Each of the above stresses is accompanied by a shear stress of 80 N/mm$^2$ and that associated with the former tensile stress tends to rotate the block anticlockwise. Determine the magnitude and direction of each of the principal stresses and the magnitude of the greatest shear stress.

***Solution*** Magnitude of major tensile, minor tensile and shear stresses are shown in Figure 13.11.

Equation of major principal stress is given by Eq. (13.18)

$$= \frac{\sigma_1 + \sigma_2}{2} + \sqrt{\left(\frac{\sigma_1 - \sigma_2}{2}\right)^2 + \tau^2}$$

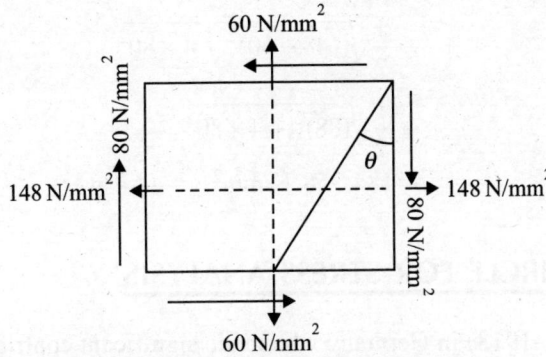

**Figure 13.11** Visual of Example 13.9.

$$= \frac{148 + 60}{2} + \sqrt{\left(\frac{148 - 60}{2}\right)^2 + 80^2}$$

$$= 104 + 91.3017$$

∴    Major principal stress = 195.3017 N/mm$^2$    (Answer)

Minor principal stress is given by Eq. (13.19)

$$= \frac{\sigma_1 + \sigma_2}{2} - \sqrt{\left(\frac{\sigma_1 - \sigma_2}{2}\right)^2 + \tau^2}$$

$$= \frac{148 + 60}{2} - \sqrt{\left(\frac{148 - 60}{2}\right)^2 + 80^2}$$

$$= 104 - 91.3017$$

$$= 12.6983 \text{ N/mm}^2    \text{(Answer)}$$

Directions of principal stresses are given by Eq. (13.20)

$$\tan 2\theta = \frac{2\tau}{\sigma_1 - \sigma_2} = \frac{2 \times 90}{148 = 60} = 2.04545$$

$$2\theta = \tan^{-1} 2.04545$$

∴    $2\theta = 63.946°$    or    $243.946°$

$\theta = 31.973°$    or    $121.973°$    (Answer)

Magnitude of greatest shear force is given by Eq. (13.23)

$$\sigma_{t \text{ max}} = \frac{1}{2}\sqrt{(\sigma_1 - \sigma_2)^2 + 4\tau^2}$$

$$= \frac{1}{2}\sqrt{(148-60)^2 + 4 \times 80^2}$$

$$= \frac{1}{2}\sqrt{(88)^2 + 4 \times 80^2}$$

$$= 184.358 \text{ N/mm}^2 \quad \text{(Answer)}$$

## 13.3  MOHR'S CIRCLE FOR STRESS ANALYSIS

It was Otto Mohr (1835–1918) in Germany who made significant contributions not only in the analysis of stress but also in the field of slope deflection of fixed beams and portal frames analysis in structural engineering. The graphical method for determination of normal, tangential stress and resultant stresses on an oblique plane by a graphical method with the help of a circle is known as famous *Mohr's circle for stress analysis*. Three different cases of drawing Mohr's circle will be presented in this section.

### 13.3.1  When a Body Subjected to Two Mutually Perpendicular Stresses

Figure 13.12 shows the Mohr's circle of stress analysis when two mutually perpendicular unequal tensile stresses act on a rectangular body. With the help of this circle, resultant stress on an oblique plane *BD* can be determined.

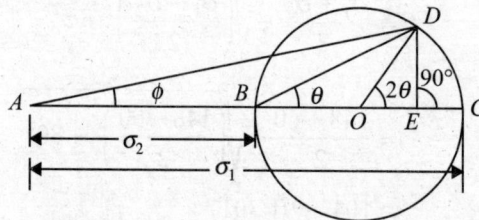

**Figure 13.12**   Mohr's circle of stress analysis.

A horizontal line through the point *A* is drawn. The lengths *AC* and *AB* are taken equal to major principal tensile stress $\sigma_1$ and minor tensile stress $\sigma_2$ to a suitable scale. Now *BC* as diameter, draw a circle with centre as *O*. Let $\theta$ be the angle made by the oblique plane with the axis of the minor tensile stress. The line *OD* is drawn in such a way that it makes the angle equal to $2\theta$ with line *OC*. From the point *D* drop a perpendicular *DE* to *AC*. Now join *AD*. Then normal and tangential stresses on the oblique plane are given by *AE* and *DE* and the resultant stress $(\sigma_R)$ on the oblique is given by *AD*.

***Proof***  Radius of the Mohr's circle $= OB = OC = OD = \dfrac{\sigma_1 - \sigma_2}{2}$

again                    $AO = AB + BO = \sigma_2 + \dfrac{\sigma_1 - \sigma_2}{2} = \dfrac{\sigma_1 + \sigma_2}{2}$

From Figure 13.11,

$$EO = OD\cos 2\theta = \frac{\sigma_1 - \sigma_2}{2}\cos 2\theta$$

But

$$AE = AO + EO$$

or

$$AE = \frac{\sigma_1 - \sigma_2}{2} + \frac{\sigma_1 - \sigma_2}{2}\cos 2\theta = \sigma_n \qquad (13.7)$$

and thus $AE$ represent the normal stress ($\sigma_n$) Eq. (13.7).

Next

$$DE = OD\sin 2\theta = \frac{\sigma_1 - \sigma_2}{2}\sin 2\theta = \sigma_t \qquad (13.8)$$

thus $DE$ represents the tangential stress ($\sigma_t$) Eq. (13.8).

## 13.3.2 When a Body Subjected to Two Mutually Perpendicular Unequal and Unlike Stresses

Mohr's circle for $\sigma_1$ as major principle tensile stress, $\sigma_2$ as minor principal compressive stress and $\theta$ as angle made by the oblique plane with the axis of minor principal stress, can also be drawn.

A horizontal line is drawn through point $A$ as shown in Figure 13.13. Cut $AB = \sigma_1$ towards positive side and $AC = \sigma_2$ towards negative side of to a suitable scale. $BC$ is bisected at $O$ and a circle is drawn $O$ as centre. Draw a line $OD$ such that it makes an angle of $2\theta$ with $OB$. From $D$, perpendicular $DE$ is drawn to $AB$. Then join $AD$ and $CD$. Now $AE$ and $DE$ are the normal ($\sigma_n$)and shear or tangential stress ($\sigma_t$) to the oblique plane. $AD$ is the resultant stress ($\sigma_R$) on the oblique plane.

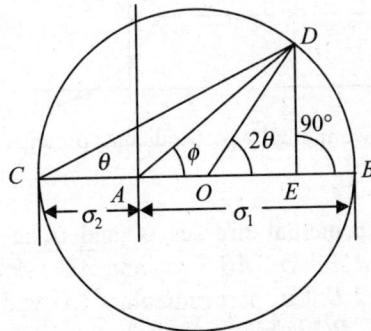

**Figure 13.13** Mohr's circle for two mutually perpendicular tensile and compressive stresses.

***Proof*** Radius of the Mohr's circle $OD = CO = OB = \dfrac{\sigma_1 + \sigma_2}{2}$

Now

$$AO = OC - AC = \frac{\sigma_1 + \sigma_2}{2} - \sigma_2 = \frac{\sigma_1 - \sigma_2}{2}$$

Again

$$\sigma_n = AE = AO + OE = \sigma_2 = \frac{\sigma_1 - \sigma_2}{2} + OD \cos 2\theta$$

or

$$\sigma_n = AE = \frac{\sigma_1 - \sigma_2}{2} + \frac{\sigma_1 + \sigma_2}{2} \cos 2\theta$$

which is the Eq. (13.7) of normal stress $\sigma_n$.
Again

$$\sigma_t = DE = OB \sin 2\theta = \frac{\sigma_1 + \sigma_2}{2} \sin 2\theta$$

which is the Eq. (13.8) of shear or tangential stress $\sigma_t$.

### 13.3.3  When a Body Subjected to Two Mutually Perpendicular Principal Stresses Accompanied by a Simple Shear

In this case, consider Figure 13.9 in Section 13.2.4 where a body is subjected to mutually perpendicular tensile stresses $\sigma_1$ and $\sigma_2$ accompanied by a simple stress $\tau$.
The Mohr's circle of above stresses is drawn in Figure 13.14.

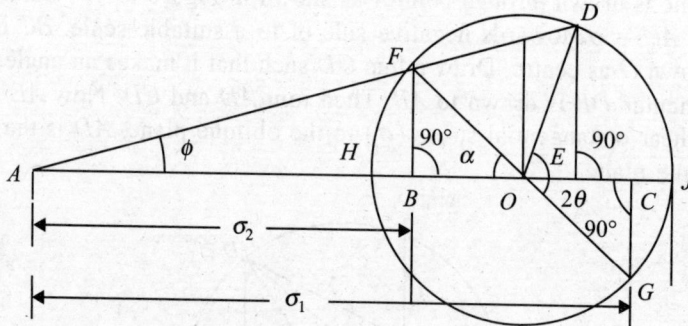

**Figure 13.14**  Mohr's circle for two mutually perpendicular principal stresses accompanied by simple shear stress.

The mutually perpendicular principal stresses $\sigma_1$ and $\sigma_2$ are shown in Figure 13.14.
In Figure 13.14, the length $AC = \sigma_1$, $AB = \sigma_2$ and $AD = \sigma_R$ to a suitable scale.
Draw perpendiculars at $C$ and $B$. Cut perpendiculars $CG$ and $BF$ equal to the tangential or shear stress $\tau$ to the same scale. $BC$ is bisected at $O$. A circle is drawn with $O$ as centre and $OF$ or $OG$ as radius. A line $OD$ is drawn in such a way that it makes the angle $2\theta$ with $OG$ as shown in Figure 13.14. Again draw another perpendicular $DE$ at $E$ to $BC$. In Figure 13.14, the lengths $AD$, $AE$ and $DE$ represent the resultant stress $\sigma_R$, $\sigma_n$ and shear or tangential stress $\sigma_t$ on the oblique plane respectively.

***Proof***  From Figure 13.14,

$$BO = \frac{1}{2} BC = \frac{1}{2}(\sigma_1 - \sigma_2)$$

$$AO = AB + BO = \sigma_2 + \frac{1}{2}(\sigma_1 - \sigma_2) = \frac{\sigma_1 + \sigma_2}{2}$$

and
$$AE = AO + OE = \frac{\sigma_1 + \sigma_2}{2} + OD \cos(2\theta - \alpha) \qquad \text{(i)}$$

Simplifying the term $OD \cos(2\theta - \alpha)$ in (i), it is obtained as:

$$AE = \sigma_n = \frac{\sigma_1 + \sigma_2}{2} + \frac{\sigma_1 - \sigma_2}{2} \cos 2\theta + \tau \sin 2\theta, \text{ which gives the Eq. (13.16).}$$

Again
$$DE = OD \sin(2\theta - \alpha) \qquad \text{(ii)}$$

Similarly, simplifying the term $OD \sin(2\alpha - \alpha)$ in (ii), it can be obtained as:

$$DE = \sigma_t = \frac{\sigma_1 - \sigma_2}{2} \sin 2\theta - \tau \cos 2\theta, \text{ which is Eq. (13.17).}$$

With the help of Mohr's circles discussed above in three different situations, solutions for normal stress $\sigma_n$, tangential stress $\sigma_t$ and resultant stress $\sigma_R$ can be obtained graphically using a suitable scale.

## 13.4   CONCLUSION

Analysis of stresses has been made when the applied force acts on an inclined or oblique section. Normal, tangential and resultant forces and its line of action or direction on an inclined plane are discussed in three different situations. Analytical equations of the above stresses in terms of principal stresses $\sigma_1$ and $\sigma_2$ and the slope of the inclined plane are derived. Mohr's circle which is the graphical method of determination of the above stresses, are presented in three different situations although analytical solutions are easy and time-saving. Few solved problems are given as to show the application of the derived equations.

## EXERCISES

**13.1** If $\sigma$ is the direct stress in one plane of a rectangular bar, show that normal stress and tangential stresses on an oblique plane which is at an angle of to the vertical plane are

$$\sigma_n = \sigma \cos^2 \theta \quad \text{and} \quad \sigma_t = \frac{\sigma}{2} \sin 2\theta$$

**13.2** Show that the maximum values of normal stress $\sigma_n$ and tangential or shear stress $\sigma_t$ in Exercise 13.1 above are $\sigma$ and $\dfrac{\sigma}{2}$ respectively.

**13.3** Describe Mohr's circle when a body subjected to two mutually perpendicular stresses. Draw the figure and prove the statements.

**13.4** A circular bar is subjected to an axial pull of $P = 200$ kN. The maximum allowable shear stress on any section is 100 N/mm$^2$. Find the diameter of the bar.

(Answer   35.682 mm)

**13.5** An axial pull of 25 kN is applied to a rectangular bar of cross-sectional area 11800 mm$^2$. Determine the normal and shear stresses on a section which is inclined at an angle of 45° with the normal cross section of the bar.

(Answer   $\sigma_n = \sigma_t = 1.0593$ N/mm$^2$)

**13.6** The stresses at a point in a bar are 140 N/mm$^2$ (tensile) and 60 N/mm$^2$ (compressive). Calculate the resultant stress in magnitude and direction on a plane inclined at 45° to the axis of the major stress. Also determine the maximum intensity of shear stress in the material at the point.

(Answer   $\sigma_R = 107.7$ N/mm$^2$, $\phi = 61°120'$, $\sigma_t = 100$ N/mm$^2$)

**13.7** At a point within a body subjected to two mutually tensile stresses of 100 N/mm$^2$ and 75 N/mm$^2$. Each of the above stresses is accompanied by a tangential stress of 75 N/mm$^2$. Determine the normal, shear and resultant stresses on an oblique plane inclined at 45° with the axis of the minor tensile stress.

(Answer   $\sigma_n = 150$ N/mm$^2$, $\sigma_t = 25$ N/mm$^2$)

**13.8** Two wooden pieces 10 cm × 10 cm in cross section are glued together along the line *EF* as shown in Figure 13.15. If the allowable shearing stress along *EF* is 0.8 N/mm$^2$, what will be the value of maximum pull *P*?

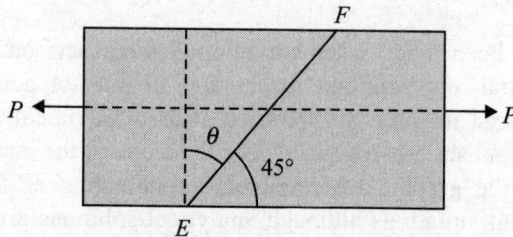

**Figure 13.15**   Visual of Example 13.8.

(Answer   *P* = 16 kN)

# CHAPTER 14

# Beams: Introduction to Bending Moment and Shear Force

## 14.1 INTRODUCTION

A *beam* is structural member supported at a few points subjected to loads transverse to its axis. In this chapter, an analysis is being presented on bending moment (BM) and shear force (SF) developed at any section of the beam in the process of transmitting the loads to the supports. The bending moment and shear force developed in the beam are, however, very much dependent on the types of beam and the types of load acting on it. Assessment of bending moment and shear force are important in designing a beam.

## 14.2 TYPES OF BEAMS

Different types of beams are as follows:

(a) Simply supported beam    (b) Cantilever beam
(c) Propped beam    (d) Beam with overhanging on one side
(e) Beam with overhanging on both sides    (f) Fixed beam
(g) Continuous beam    (h) Hinged beam

Different types of beams are shown symbolically as a line *AB* in Figure 14.1. Figure 14.1(a) represents a simply supported beam with two reactions at supports *A* and *B*. Figure 14.2(b) is the cantilever beam fixed at end *A* without support at *B*. At the fixed end a vertical reaction and a moment $M_A$ and its direction is shown. In propped cantilever beam shown in Figure 14.1(c), a support or prop is provided at point *B*. End reactions and end moments are the same as of cantilever. Beam with overhanging on one side towards *B* and two reactions at *A* and *B* is shown in Figure 14.1(d). Similarly beams with two overhanging on two sides at *A* and *B* is presented in Figure 14.1(e). A fixed beam *AB*, fixed at both ends *A* and *B* is shown with two reactions and two moments in Figure 14.1(f). A continuous beam fixed at one end and the number of spans with supports is shown in Figure 14.1(g). Fixed end provides end reaction and moment as shown in the figure. Figure 14.1(h) shows a hinged beam where a hinge is used to connect two segments of a beam.

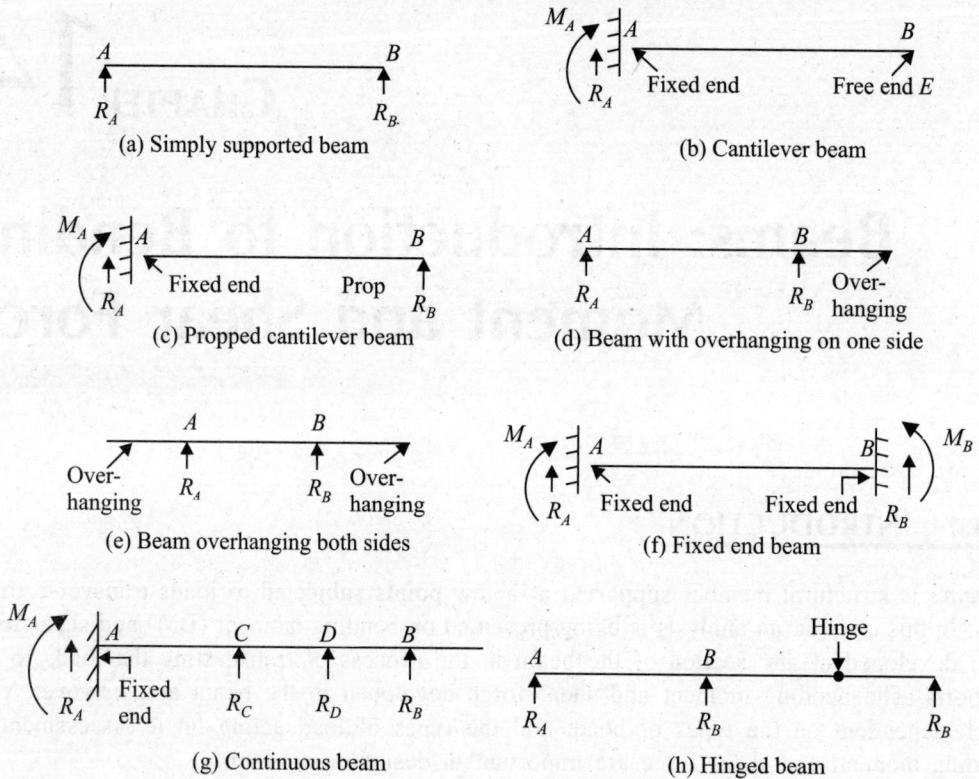

Figure 14.1   Different types of beam.

## 14.3   TYPES OF LOADING

Beams may be subjected to various types of load like:

(a) Concentrated or point load
(b) Uniformly or varying distributed load.

Different types of loading like concentrated or point loads, uniformly distributed per unit length, uniformly varying load per unit length are shown in Figure 14.2(a), (b) and (c) respectively.

## 14.4   SHEAR FORCE AND BENDING MOMENT

Shear force (SF) at a section of beam or structural member can be defined as the force that is trying to shear off the section and is obtained as the sum of the normal forces on side of the section.

Bending moment (BM) at a section of a beam or structural member is the moment that is trying to bend the member and is obtained as the algebraic sum of the moments of all forces about the section acting either to the left or to the right of the section.

(a) Beam with one or more point loads    (b) Beam with uniformly distributed load

(c) Beam with uniformly distributed load

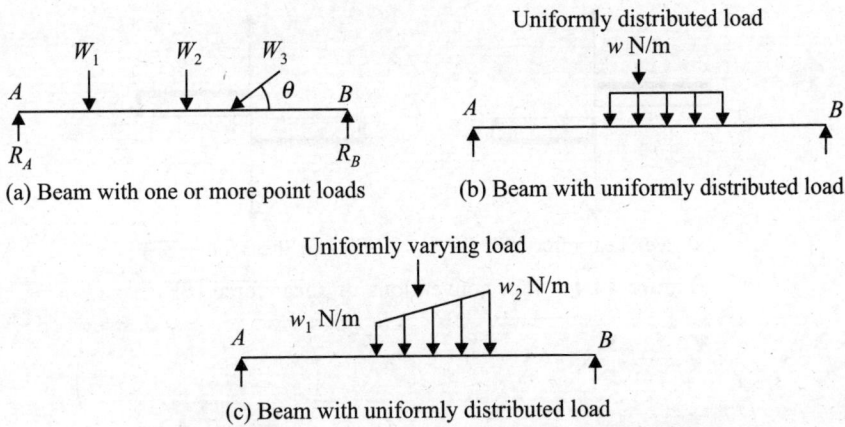

**Figure 14.2**   Types of loading on beam.

Figure 14.3 shows how a beam shears off at a section $C$ due to shear force (SF) and tends to bend the beam at point $C$. This illustration clearly represents the concept of action of SF and BM on any beam.

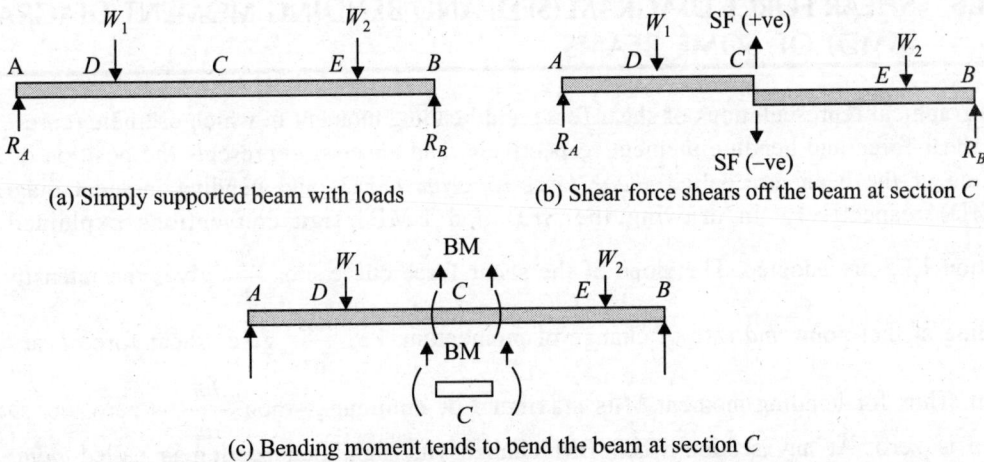

(a) Simply supported beam with loads    (b) Shear force shears off the beam at section $C$

(c) Bending moment tends to bend the beam at section $C$

**Figure 14.3**   SF shears off and BM tends to bend the beam at $C$.

## 14.5   SIGN CONVENTION OF SF AND BM

The most general sign conventions for shear force (SF) and bending moment (BM) are as follows. The shear force that tends to move the left portion upward as shown in Figure 14.4(a) is called *positive shear force*. If the left portion tends to move downwards, as shown in Figure 14.4(b), it is called *negative shear force*.

The bending moment that tends to sag the beam is taken as positive bending moment. It is also called *sagging moment* and is shown in Figure 14.5(a). On the other hand, if the bending moment to hog, as shown in Figure 14.5(b), it is negative and is shown in the same figure.

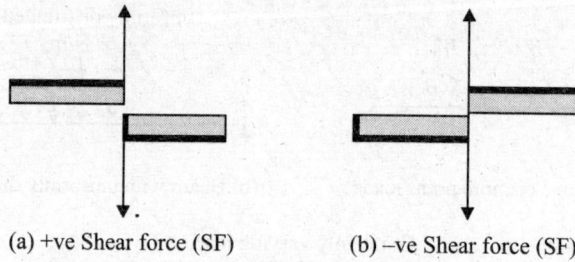

(a) +ve Shear force (SF)          (b) −ve Shear force (SF)

**Figure 14.4**    Sign conventions of shear force (SF).

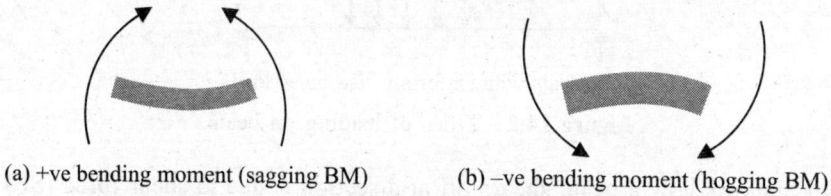

(a) +ve bending moment (sagging BM)        (b) −ve bending moment (hogging BM)

**Figure 14.5**    Sign conventions of bending moment.

## 14.6    SHEAR FORCE DIAGRAM (SFD) AND BENDING MOMENT DIAGRAM (BMD) OF SOME BEAMS

The graphical representations of shear force and bending moment in which ordinate represents the shear force and bending moment respectively, and abscissa represents the position of the section of the beam are called *shear force diagram* (SFD) and bending moment diagram (BMD) respectively. In drawing the SFD and BMD, sign conventions explained in Section 13.5 are adopted. The slope of the shear force curve, i.e. $\dfrac{dF}{dx}$ gives the intensity of loading at that point and rate of change of momentum, i.e. $\dfrac{dM}{dx}$ gives shear force $F$ at that point. Thus for bending moment $M$ is maximum or minimum, when $\dfrac{dM}{dx}$ is zero, i.e. shear force is zero. At any section when BM changes its sign, that section is called *point of contraflexure* and obviously BM at that section is zero. These SFD and BMD are located below the load diagram. Shear force diagram (SFD) and bending moment diagram (BMD) for some simple beams have been shown in this section.

### 14.6.1    When Simply Supported Beam with One Concentrated Load

Figure 14.6(a) shows that the concentrated load $W$ is placed at point $C$ which is at a distance $a$ from support point $A$ and $b$ from support $B$. To determine reaction $R_A$ take moment about $B$ of all the forces:

$$R_A \times L = W \times b + R_B \times 0$$

$\therefore$
$$R_A = \frac{Wb}{L}$$

Similarly

$$R_B = \frac{Wa}{L}$$

Now in the left portion $AC$ of the beam, shear force $F$ is constant and equal to $R_A = \dfrac{Wb}{L}$ and is positive as seen from SFD diagram 14.6(b). Similarly in the right portion $BC$ of the beam, shear force $F$ is constant and equal to $-R_B = -\dfrac{Wa}{L}$ and is negative as seen in the same SFD diagram.

Next, considering the moment $M$ in the portion $AC$ at a distance $x$ from $A$,

$$M = R_A x = \frac{Wb}{L} x$$

When $x = 0$, $M = 0$ and at $x = a$,

$$M = \frac{Wab}{L} \tag{14.1}$$

Again considering the moment $M$ in the portion $BC$ at a distance $x$ from $B$,

$$M = R_B x = \frac{Wa}{L} x$$

When $x = 0$, $M = 0$, and $x = b$,

$$M = \frac{Wab}{L}$$

Now the bending moment diagram (BMD) is drawn with $M = 0$ at $A$, $M = \dfrac{Wab}{L}$ at $C$ and $M = 0$ at $B$ is drawn in Figure 14.6(c).

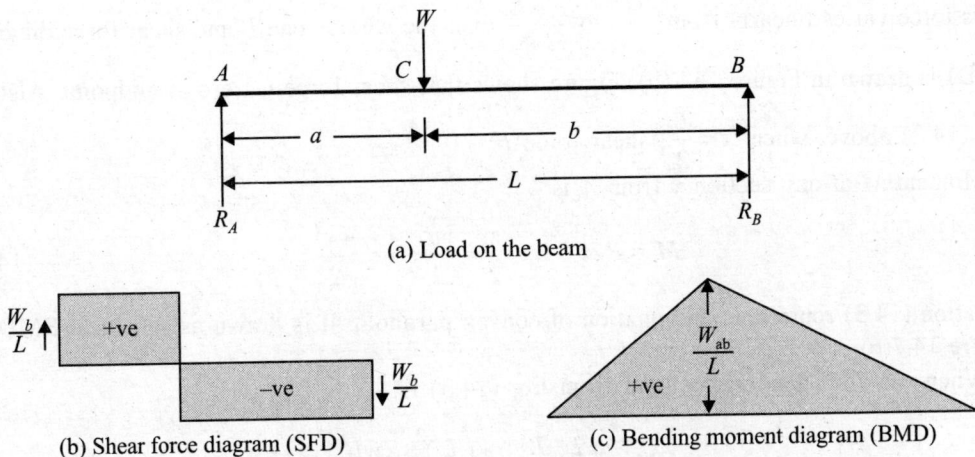

(a) Load on the beam

(b) Shear force diagram (SFD)

(c) Bending moment diagram (BMD)

**Figure 14.6**   SFD and BMD of simply supported beam with a point load $W$.

When $a = b = \dfrac{L}{2}$, i.e. the point load $W$ is at the centre of the beam, Eq. (14.1) becomes,

$$M = \frac{Wab}{L} = \frac{W \cdot \dfrac{L}{2} \cdot \dfrac{L}{2}}{L} = \frac{WL}{4}$$

Thus maximum bending moment occurs when the load $W$ is at the centre $M_{max}$ is:

$$M_{max} = \frac{WL}{4}$$

### 14.6.2   When Simply Supported Beam with Uniformly Distributed Load (UDL)

Figure 14.7(a) shows the a simply supported beam $AB$ uniformly distributed load (UDL) of intensity $w$/unit length. Since the load is uniformly distributed over the whole span of $L$, reactions at supports $A$ and $B$ be equal, i.e. equal to the half of the total load at each support.
  Hence

$$R_A = R_B = \frac{wL}{2}$$

Shear force $F$ at a distance $x$ from support $A$ is:

$$F = R_A - wx = \frac{wL}{2} - wx \tag{14.2}$$

Equation (14.2) gives a linear variation. When $x = 0$,

$$F = R_A = \frac{wL}{2}$$

and when $x = L$

$$F = R_B = -\frac{wL}{2}$$

Thus force varies linearly from $\dfrac{wL}{2}$ to $-\dfrac{wL}{2}$ over the whole span $L$ and shear force diagram (SFD) is drawn in Figure 14.7(b). Figure shows that shear force is zero at midpoint. Also in Eq. (14.2) above, when $x = \dfrac{L}{2}$, shear force $F = 0$.
  Moment $M$ at any section $x$ from $A$ is:

$$M = R_A x - wx \cdot \frac{x}{2} = \frac{wL}{2} \cdot x - \frac{wx^2}{2} \tag{14.3}$$

Equation (14.3) represents an equation of convex parabola. It is drawn as shown in BMD in Figure 14.7(c).
  When $x = L/2$, bending moment from Eq. (14.3) is:

$$M = \frac{wL}{2} \cdot \frac{L}{2} - \frac{w}{2}\left(\frac{L}{2}\right)^2 = \frac{wL^2}{8} \tag{14.4}$$

(a) Load on the beam

(b) Shear force diagram (SFD)

(c) Bending moment diagram (BMD)

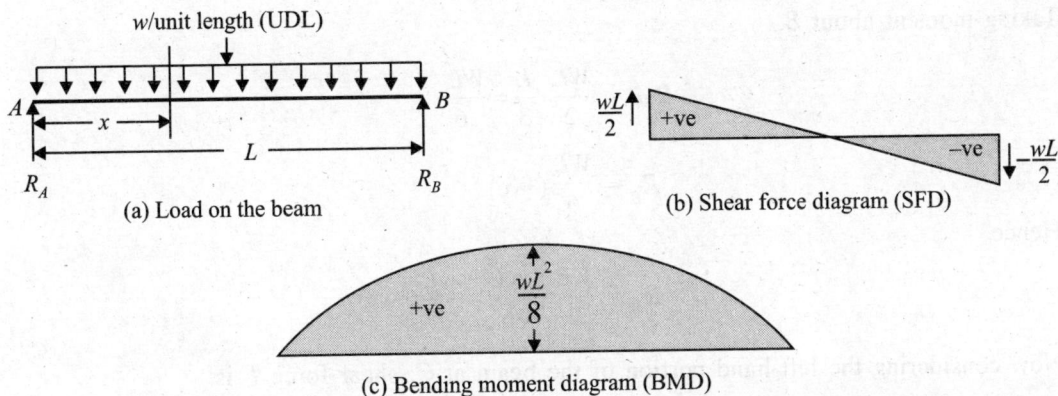

**Figure 14.7**   SFD and BMD of simply supported beam with UDL $w$/unit length.

Again differentiating $M$ in Eq. (14.3) with respect to $x$ and equating to zero for maximum,

$$\frac{dM}{dx} = \frac{wL}{2} - \frac{2wx}{2} = 0$$

or

$$x = \frac{L}{2}$$

Substituting this $x$ values in Eq. (14.3) for maximum bending moment,

$$M_{max} = \frac{wL}{2} \cdot \frac{L}{2} - \frac{w}{2} \cdot \left(\frac{L}{2}\right)^2 = \frac{wL^2}{8} \tag{14.5}$$

Equation of maximum bending moment given by Eq. (14.5) is same with Eq. (14.4). Thus it can be concluded that maximum bending moment occurred at midpoint of the span $L$.

## 14.6.3  When Simply Supported Beam with Uniformly Varying Load (UVL)

Figure 14.8(a) shows the simply supported beam $AB$ of span of $L$ with uniformly varying load (UVL) on it. To draw the SFD and BMD of the beam, consider a distance $x$ from support $A$. The uniformly varying load (UVL) is $w$/unit length.

Hence total load in this portion $AC = \dfrac{1}{2}$(load at $C$ + load at $A$) $\times x = \dfrac{1}{2}\left(\dfrac{wx}{L} + 0 \times x\right) = \dfrac{wx^2}{2L}$

The CG of this load from $C$ is at a distance of $\dfrac{x}{3}$

Similarly total load on the beam $= \dfrac{wL^2}{2L} = \dfrac{wL}{2}$

The CG of this load from $B$ is at a distance of $\dfrac{L}{3}$

Taking moment about $B$,

$$R_A L = \frac{WL}{2} \cdot \frac{L}{3} = \frac{WL^2}{6}$$

$$R_A = \frac{WL}{6} \text{ (+ve)}$$

Hence

$$R_B = \frac{wL}{2} - \frac{wL}{6} = \frac{wL}{3} \text{ (−ve)}$$

Now considering the left-hand portion of the beam at $C$, shear force $F$ is:

$$F = R_A - \frac{wx^2}{2L} = \frac{wL}{6} - \frac{wx^2}{2L} \qquad (14.6)$$

Equation (14.6) represents a convex parabola.

At

$$x = 0, \ F_A = \frac{wL}{6}$$

At

$$x = L, \ F_B = \frac{wL}{6} - \frac{wL^2}{2L} = -\frac{wL}{3}$$

The section for zero shear force is obtained by substituting $F = 0$ in Eq. (14.6), i.e.

$$0 = \frac{wL}{6} - \frac{wx^2}{2L}$$

or

$$\frac{wx^2}{2L} = \frac{wL}{6}$$

Solving for $x$,

$$x = \frac{L}{\sqrt{3}}$$

Equation (14.6) is plotted for different values of $x$ to show the convex parabola in Figure 14.8(b).

Taking moment at $C$,

$$M = R_A x - \text{Load on } AC \times \frac{x}{3}$$

$$M = \frac{wL}{6} x - \frac{wx^2}{2L} \cdot \frac{x}{3} = \frac{wLx}{6} - \frac{wx^3}{6L} \qquad (14.7)$$

Equation (14.7) represents a cubic parabola.
At $x = 0, M_A = 0$ and at $x = L \cdot M_B = 0$

We know that moment is the maximum where shear force $F$ is zero, i.e. at $x = \dfrac{L}{\sqrt{3}}$ Substituting this value of $x$ in Eq. (14.7),

$$M_{\max} = \frac{wL}{6} \cdot \frac{L}{\sqrt{3}} - \frac{w}{6L} \cdot \left(\frac{L}{\sqrt{3}}\right)^3 = \frac{wL^2}{9\sqrt{3}} = 0.06415 \, wL^2$$

Thus BMD is drawn with the help of Eq. (14.7) in Figure 14.8(c) where maximum moment is shown at a $x = \dfrac{L}{\sqrt{3}}$ from $A$.

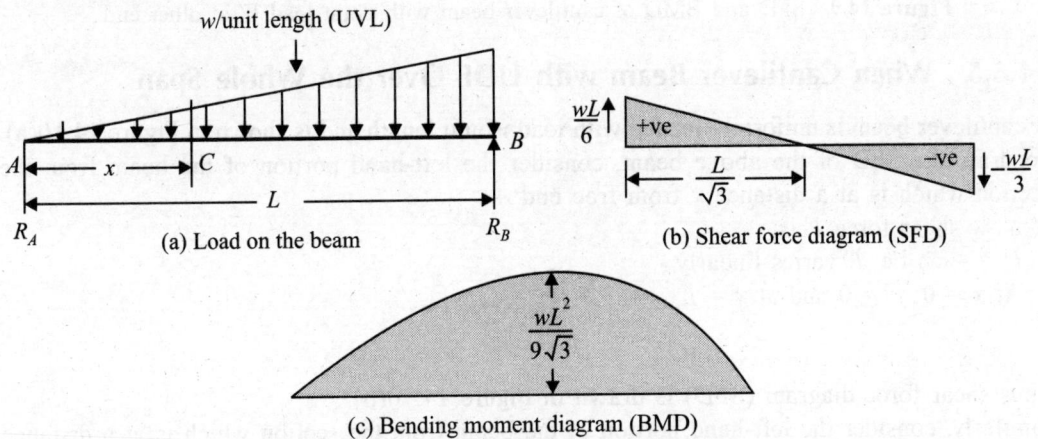

(a) Load on the beam

(b) Shear force diagram (SFD)

(c) Bending moment diagram (BMD)

**Figure 14.8**   SFD and BMD of simply supported beam with UVL $W$/unit length.

## 14.6.4  When Cantilever Beam with One Concentrated Load at Free End

Figure 14.9(a) is the cantilever beam with concentrated load at free end. To draw the shear force diagram (SFD) of this beam:

From left hand segment of the beam, shear force,

$$F = -W$$

Hence shear force remains constant throughout and does not vary with $x$. Thus shear force diagram is shown in Figure 14.9(b).

For bending moment diagram (BMD),

$$M = -Wx \tag{14.8}$$

Equation (14.8) shows that bending moment varies linearly
At $x = 0$, $M_A = 0$ and at $x = L$,

$$M_B = -WL$$

Thus bending moment diagram is drawn in Figure 14.9(c).

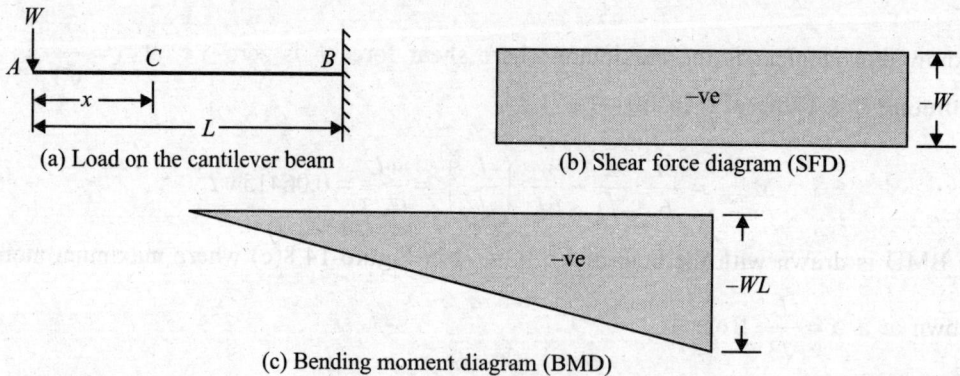

(a) Load on the cantilever beam    (b) Shear force diagram (SFD)

(c) Bending moment diagram (BMD)

**Figure 14.9** SFD and BMD of cantilever beam with point load $W$ at other end.

## 14.6.5 When Cantilever Beam with UDL Over the Whole Span

A cantilever beam is uniformly loaded with load $w$/unit length and is shown in Figure 14.10(a). To draw the SFD of the above beam, consider the left-hand portion of the beam from the section which is at a distance $x$ from free end $A$.

Now shear force $F$ is:

$F = -wx$, i.e. $F$ varies linearly.

At $x = 0$, $F = 0$ and at $x = L$

$$F = -WL$$

Thus shear force diagram (SFD) is drawn in Figure 14.10(b).

Similarly, consider the left-hand portion of the beam from the section which is at a distance $x$ from free end $A$ for drawing the BMD.

Bending moment

$$M = -wx \cdot \frac{x}{2} = -\frac{wx^2}{2} \tag{14.9}$$

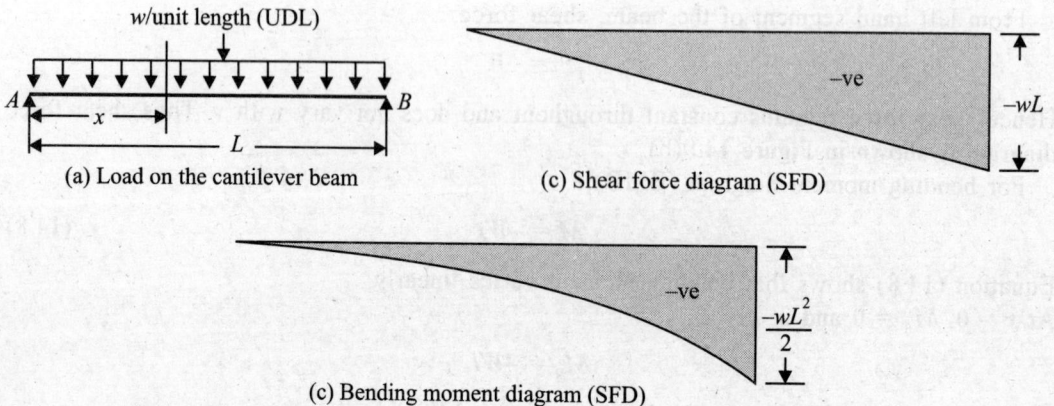

(a) Load on the cantilever beam    (c) Shear force diagram (SFD)

(c) Bending moment diagram (SFD)

**Figure 4.10** SFD and BMD of cantilever beam with UDL in entire span.

Equation (14.9) is parabolic. It gives a concave parabola. The values at $x$ and $L$ are:

At $x = 0$, $M_A = 0$ and at $x = L$

$$M_B = -\frac{wL^2}{2}$$

Thus the bending moment diagram of concave parabola is shown in Figure 14.10(c).

### 14.6.6    When Cantilever Beam with Uniformly Varying Load (UVL) Over the Whole Span

A cantilever beam with uniformly varying load (UVL) over the whole span is shown in Figure 14.11(a). To draw the SFD, consider the intensity of load at a distance $x$ from the free end. If $w_x$ is intensity of load at $x$, then:

$$\frac{w_x}{x} = \frac{w}{L}$$

$$\therefore \qquad w_x = \frac{wx}{L}$$

$\therefore$   Shear force   $\qquad F = -\frac{1}{2} \cdot \frac{wx}{L} \cdot x = -\frac{wx^2}{2L}$   (14.10)

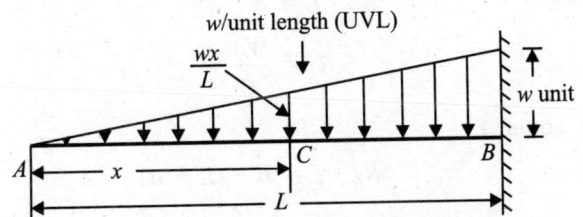

(a) Load on the cantilever beam

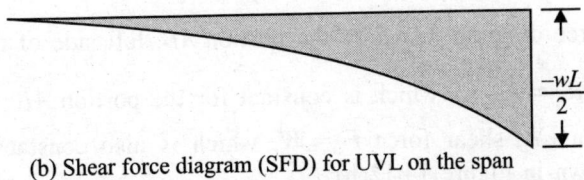

(b) Shear force diagram (SFD) for UVL on the span

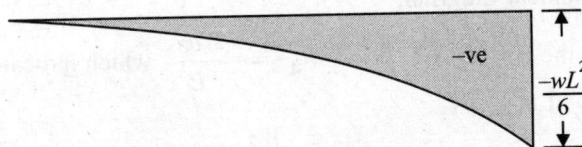

(c) Bending moment diagram (BMD)

**Figure 14.11**   SFD and BMD of cantilever beam with UVL on entire span.

Equation (14.10) gives the parabola of a concave curve.
At $x = 0$, $F_A = 0$ and at $x = L$

$$F_B = -\frac{wL}{2}$$

Thus the shear force diagram of concave parabola is drawn in Figure 14.11(b).
   To draw the BMD, bending moment at a distance $x$ from the free end is:

$$M = -\frac{1}{2}\frac{wx^2}{L}\cdot\frac{x}{3} = -\frac{wx^3}{6L} \qquad (14.11)$$

Equation (14.11) shows that moment $M$ varies with $x$ to form a concave parabola.
   When $x = 0$, $M_A = 0$ and when $x = L$

$$M_B = M_B = -\frac{wL^2}{6}$$

BMD is drawn in Figure 14.11(c).

### 14.6.7   When Overhanging Beam with a Concentrated Load at Free End

The overhanging beam $ABC$ with a concentrated load $W$ at free end $C$ is shown in Figure 14.12(a). To draw its BMD, it is required to calculate the two reactions $R_A$ and $R_B$.
   Taking moment about $B$,

$$R_A \cdot L = W \cdot a$$

$\therefore$
$$R_A = \frac{Wa}{L}$$

Again taking moment about $A$,

$$R_B \cdot L = W \cdot (L + a)$$

$\therefore$
$$R_B = W\left(1 + \frac{a}{L}\right)$$

To draw the shear force diagram, consider the portion $AB$, left side of the section.

   Shear force $F = -R_A = -\dfrac{Wa}{L}$, which is constant for the portion $AB$.

   Considering portion $CB$, shear force $F = W$, which is also constant. Thus shear force diagram (SFD) is drawn in Figure 14.12(b).
   To draw bending moment diagram,

   Bending moment in the portion $AB$, $M = -R_A x = -\dfrac{Wax}{L}$, which indicates $M$ varies linearly.
   At $x = 0$, $M_A = 0$ and at $x = L$

$$M_B = -Wa$$

Bending moment in the portion $CB$,

$$M = -Wx$$

At $x = 0$, $M_C = 0$ and at $x = a$

$$M_B = -Wa$$

With these calculated values at $A$, $B$ and $C$, bending moment diagram (BMD) is drawn in 14.12(c).

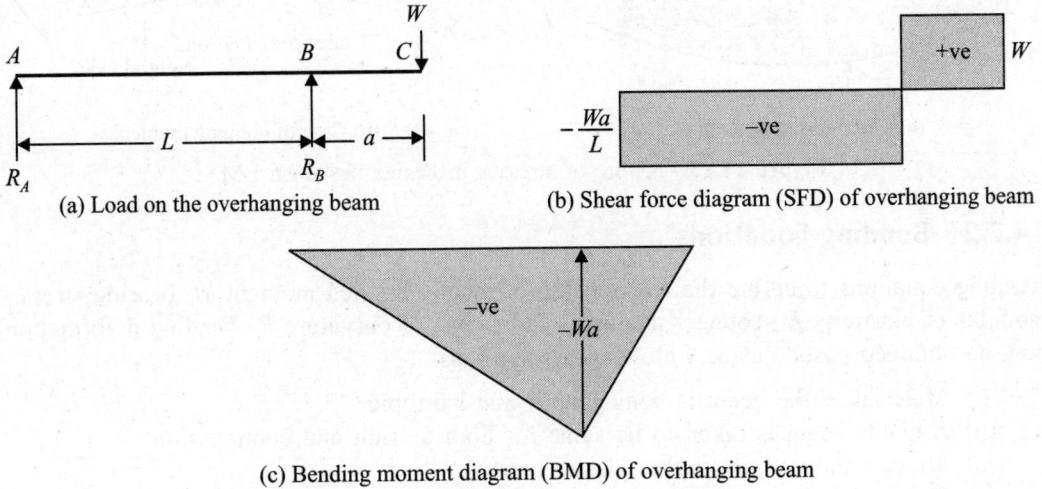

(a) Load on the overhanging beam

(b) Shear force diagram (SFD) of overhanging beam

(c) Bending moment diagram (BMD) of overhanging beam

**Figure 14.12**    SFD and BMD of overhanging beam with concentrated load at the free end.

## 14.7    SIMPLE BENDING THEORY

The bending moment and shear force on beam under different loading patterns vary from section to section of the beam. Stresses in the beam develop to resist the bending moments and shear forces. The theory in which stresses developed in the section of beam is called *simple bending theory*. Analysis of this simple bending theory will be discussed in this section.

### 14.7.1    Nature of Stresses in Beam

It is already shown in Figure 14.5 that beam's sagging or hogging moment depends upon the nature of bending moment. Figure 14.13 is drawn to explain this sagging and hogging of beams in details.

It is seen in Figure 14.13(a) that beam fibres at top side are compressed while fibres in bottom side are stretched. It indicates that the material of the beam at top side is subjected to compressive stress and to tensile stress in bottom side. But in case of hogging, as shown in Figure 14.13(b), the nature of bending stresses are exactly opposite, i.e. tensile stress in top side and compressive stress in bottom side. Thus bending stress in a beam depending on loading varies from tensile at one edge to compressive on the other edge. Bending stress somewhere between the two edges must be zero. The layer where bending stress is zero is called *neutral layer* and trace of this neutral layer is called *neutral axis* which is marked in both Figures 14.13(a) and 14.13(b).

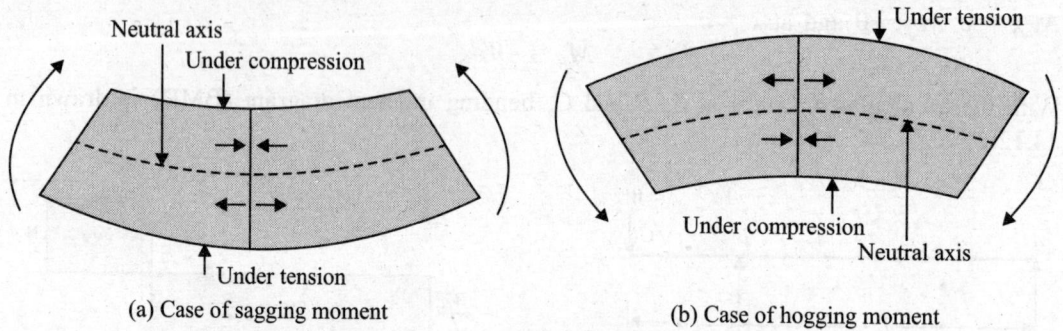

(a) Case of sagging moment          (b) Case of hogging moment

**Figure 14.13**   Nature of stresses in beams based on BM.

### 14.7.2   Bending Equations

Bending equations to define the relationship of among applied moment $M$, bearing stress $f$, modulus of elasticity $E$ (Young's modulus) and radius of curvature $R$ (bending deformation) may be obtained based on the following assumptions:

   (i) Material of the beam is homogenous and isotropic
  (ii) $E$ of the beam is taken to be same for both tension and compression
 (iii) Stresses induced are within elastic limit
 (iv) Plane section remains plane even after deformation
  (v) Every layer of initially straight beam is free to expand or contract
 (vi) Radius of curvature is quite large compared to depth of the beam

### 14.7.3   Relation among Bending Stress *f*, Radius of Curvature *R* and Modulus of Elasticity *E*

A straight elemental strip $ABCD$ of length $AB$ of a beam is shown in Figure 14.14. Let $EF$ be the neutral axis. Another layer $GH$ is considered at a depth of $y$ below the neutral layer. $CD$ is the top layer and $AB$ is the bottom layer. Initially before bending, $AB = GH = EF = CD$.

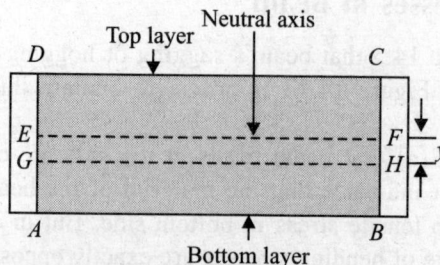

**Figure 14.14**   A straight elemental length $AB$ of the beam.

Figure 14.15 is drawn after bending the elemental strip $ABCD$ of the beam in Figure 14.14 and points $A$, $B$, $C$, $D$, $E$, $F$, $G$ and $H$ take the position $A'$, $B'$, $C'$, $D'$, $E'$, $F'$, $G'$ and $H'$ respectively. Let $R$ be the radius of curvature and $\phi$ be the angle subtended by $A'D'$ and $B'C'$ at the centre of radius of curvature $O$.

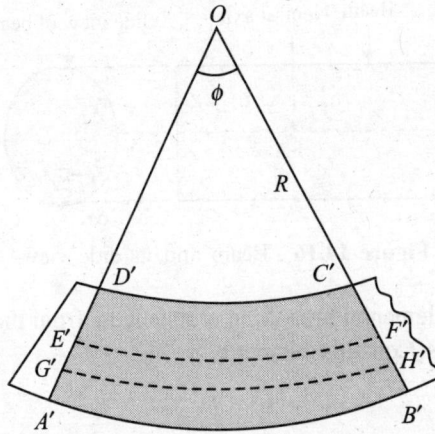

**Figure 14.15**  Elemental beam $AB$ after bending.

Now

$$EF = E'F' = R\phi$$

$\because$ $EF$ is the neutral axis and angle $\phi$ is small.

$$\text{Strain in the layer } GH = \frac{\text{Final length} - \text{Initial length}}{\text{Initial length}} = \frac{G'H' - GH}{GH}$$

But
$$GH = EF = R\phi$$

and
$$G'H' = (R + y)\phi$$

$\therefore$
$$\text{Strain in the layer } GH = \frac{(R + y)\phi - R\phi}{R\phi}$$

$$= \frac{y}{R} \tag{14.12}$$

Since strain $GH$ is due to tensile stress $f$,

$$\text{Strain in } GH = \frac{f}{E} \tag{14.13}$$

Equating (14.12) and (14.13)

$$\frac{y}{R} = \frac{f}{E}$$

or
$$\frac{f}{y} = \frac{E}{R} \tag{14.14}$$

## 14.7.4  Relation among *M, R, E, f, y* and Moment of Inertia *I*

In order to establish the relationship among the above six variables, a beam and its sectional view have been shown in Figure 14.16.

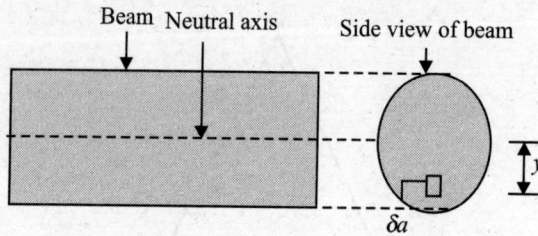

**Figure 14.16** Beam and its side view.

In the sectional view, an elemental area $\delta a$ at a distance $y$ from the neutral axis is considered. From Eq. (14.14), the stress $f$ on this element,

$$f = \frac{E}{R} y \tag{14.15}$$

Hence force on this element = stress × area = $f \cdot \delta a = \frac{E}{R} y \delta a$

Moment of resistance of this force about the neutral axis $f \cdot \delta a = \frac{E}{R} y \delta a \cdot y$

$$= \frac{E}{R} y^2 \delta a$$

∴ Total moment $M'$ resisted by the section is given by:

$$M' = \frac{E}{R} \sum y^2 \cdot \delta a$$

But the term $\sum y^2 \cdot \delta a$ represents the moment of inertia $I$ (second moment of area) by definition.

∴ $$M' = \frac{E}{R} I$$

Applying condition of equilibrium, $M' = M$ (applied moment)

∴ $$M = \frac{E}{R} I$$

or $$\frac{M}{I} = \frac{E}{R} \tag{14.16}$$

From Eqs. (14.14) and (14.15), it is obtained:

$$\frac{M}{I} = \frac{f}{y} = \frac{E}{R} \tag{14.17}$$

Equations (14.15), (14.16) and (14.17) are called the *bending equations*.

## 14.8    LOCATION OF NEUTRAL AXIS

Consider the elemental area $\delta a$ of Figure 14.16 at a distance $y$ from the neutral axis (NA). If the bending stress on the area is $\delta a$, then force on it $= f \cdot \delta a$. But Eq. (14.15) is:

$$f = \frac{E}{R} y$$

Hence force on the element $= \dfrac{E}{R} y \cdot \delta a$

∴ Total horizontal force on the beam is:

$$\sum \frac{E}{R} y \cdot \delta a = \frac{E}{R} \sum y \cdot \delta a$$

This horizontal force is in equilibrium as there is no other horizontal force and for equilibrium condition,

$$\frac{E}{R} \sum y \cdot \delta a = 0 \tag{14.18}$$

Now $\dfrac{E}{R}$ in Eq. (14.18) is not zero, hence $\sum y \cdot \delta a$ must be zero.

If $A$ is the total area of the cross section, then $\sum \dfrac{y \delta a}{A} = 0$

In Eq. (14.18), the term $\sum y \cdot \delta a$ is the moment of area about the neutral axis and $\dfrac{\sum y \delta a}{A}$ should be the distance of centroid of the area from the neutral axis.

∴ $\sum \dfrac{y \delta a}{A} = 0$ Means that neutral axis coincides with the centroid of the cross section.

## 14.9    MAXIMUM MOMENT OF A SECTION

Equation (14.17) for bending stress is:

$$\frac{M}{I} = \frac{f}{y} = \frac{E}{R}$$

or

$$f = \frac{M}{I} y$$

This bending stress $f$ is the maximum when $y$ is the maximum. It means $f$ is the maximum in the extreme fibre. Let the distance of extreme fibre be $y_{max}$,

Maximum bending stress is

$$f_{max} = \frac{M}{I} y_{max}$$

or
$$M = \left(\frac{I}{y_{max}}\right) f_{max} \qquad (14.19)$$

The term $\left(\dfrac{I}{y_{max}}\right)$ in Eq. (14.19) represents the properties of the section and, therefore, this term is called *modulus of section* or *section modulus* which is denoted by Z. The section modulus Z has the unit of (length)$^3$.

Hence section modulus

$$Z = \frac{I}{y_{max}} \qquad (14.20)$$

Equation (14.19) becomes

$$M = f_{max} Z \qquad (14.21)$$

Equation (14.21) gives the permissible maximum moment of the section and hence it is written as

$$M_{max} = F_{max} Z$$

Thus with the help of section modulus, Z is a measure of moment, of which the resistance can be predicted.

## 14.10   SECTION MODULI OF STANDARD SECTION

Equations of section moduli Z of different sections are derived in the following subsections.

### 14.10.1   Section Modulus of Rectangular Section

As seen from Figure 14.17 for rectangular section,

$$y_{max} = \frac{d}{2}$$

Moment of Inertia for rectangular section about the neutral axis (NA) is:

$$I = \frac{1}{12}bd^3$$

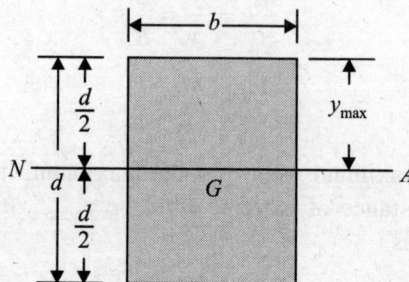

**Figure 14.17**   Section modulus of rectangular section.

Hence section modulus

$$Z = \frac{I}{y_{max}} = \frac{\frac{1}{12}bd^3}{\frac{d}{2}} = \frac{1}{6}bd^2$$

## 14.10.2  Section Modulus of Triangular Section

Moment of inertia $I$ of the triangular section about NA, shown in Figure 14.18, is:

$$I = \frac{1}{36}bh^3 \quad \text{and} \quad y_{max} = \frac{2}{3}h$$

∴ Section modulus

$$Z = \frac{I}{y_{max}} = \frac{\frac{1}{36}bh^3}{\frac{2}{3}h} = \frac{1}{24}bh^2$$

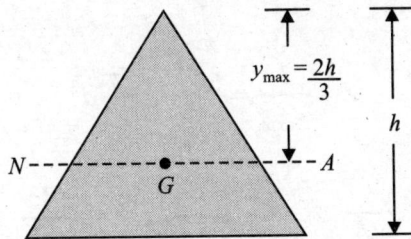

**Figure 14.18**  Section modulus of triangular section.

## 14.10.3  Section Modulus of Hollow Rectangular Section

A symmetrical hollow opening rectangular beam is shown in Figure 14.19. Moment of inertia $I$ is:

$$I = \frac{BD^3}{12} - \frac{bd^3}{12} = \frac{1}{12}(BD^3 - bd^3)$$

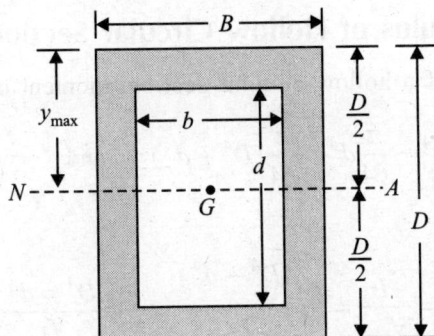

**Figure 14.19**  Section modulus of hollow rectangular section.

As shown in Figure 14.19

$$y_{max} = \frac{D}{2}$$

∴ Section modulus

$$Z = \frac{I}{y_{max}} = \frac{\frac{1}{12}(BD^3 - bd^3)}{\frac{D}{2}} = \frac{1}{6}\left(\frac{BD^3 - bd^3}{D}\right)$$

### 14.10.4   Section Modulus of Circular Section

Considering Figure 14.20 of circular section, moment of inertia is:

$$I = \frac{\pi}{64}d^4 \quad \text{and} \quad y_{max} = \frac{d}{2}$$

∴ Section modulus $Z$ is

$$Z = \frac{I}{y_{max}} = \frac{\frac{\pi}{64}d^4}{\frac{d}{2}} = \frac{\pi}{32}d^3$$

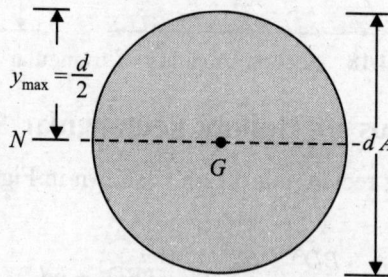

**Figure 14.20**   Section modulus of circular section.

### 14.10.5   Section Modulus of Hollow Circular Section

Considering Figure 14.21 of a hollow circular section, moment of inertia is:

$$I = \frac{\pi}{64}D^4 - \frac{\pi}{64}d^4 = \frac{\pi}{64}(D^4 - d^4) \quad \text{and} \quad y_{max} = \frac{D}{2}$$

∴ Section modulus

$$Z = \frac{I}{y_{max}} = \frac{\frac{\pi}{64}(D^4 - d^4)}{\frac{D}{2}} = \frac{\pi}{32}\left(\frac{D^4 = d^4}{D}\right)$$

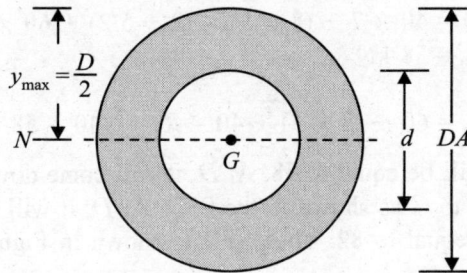

**Figure 14.21**    Section modulus of hollow circular section.

***EXAMPLE 14.1***    A cantilever beam of length 3 m carries a UDL of 20 N/m throughout the length. Calculate the bending moment of the beam near the fixed end. What is the SF at this point?

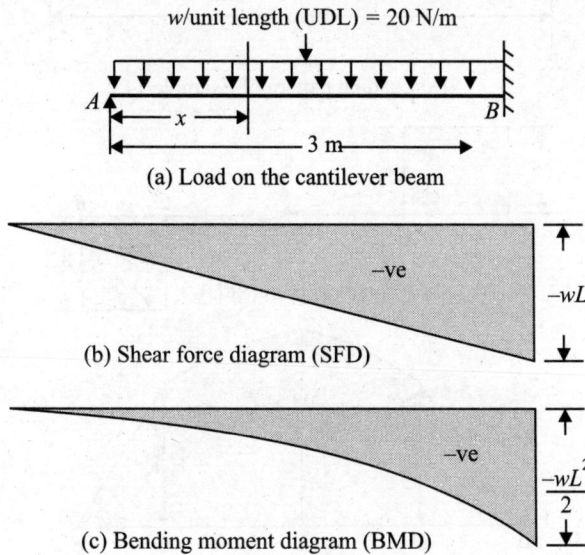

(a) Load on the cantilever beam

(b) Shear force diagram (SFD)

(c) Bending moment diagram (BMD)

**Figure 14.22**    SFD and BMD of Example 14.1.

***Solution***    SFD and BMD of the cantilever beam of 3 m are drawn and shown below.

From Figure (14.22) BM near the fixed end $= -\dfrac{wL^2}{2} = -\dfrac{20 \times 3^2}{2} = 90$ Nm.     (Answer)

Shear force near the fixed end $= -WL = -20 \times 3 = -60$ N.     (Answer)

***EXAMPLE 14.2***    A simply supported beam $AB$ has a span of 10 m. Two concentrated loads 60 kN and 40 kN act at $C$ and $D$ at a distance of 2 m and 7 m from support $B$. Further, the part $CD$ is loaded by 8 kN/m. Analyze and draw BMD and SFD.

***Solution***    The loading pattern on the beam $AB$ as given in Example 14.2 is shown in Figure 14.23(a). To analyze for SFD: taking moment about $B$,

$$R_A \times 10 = 40 \times 7 + (8 \times 5) \times (2 + 5/2) + 60 \times 2$$
$$R_A = 58 \text{ kN}$$

Hence

$$R_B = 60 + (8 \times 5) + 40 - R_A = 140 - 58 = 82 \text{ kN}$$

Shear force from $A$ to $D$ will be equal to 58. At $D$, it will come down vertically 40 and from $D$ to $C$ it varies linearly $D$ to $C$ as shown in the SFD. At $D$, it will come down further down and will be constant to $B$ equal to 82. Thus SFD is shown in Figure 14.23(b).

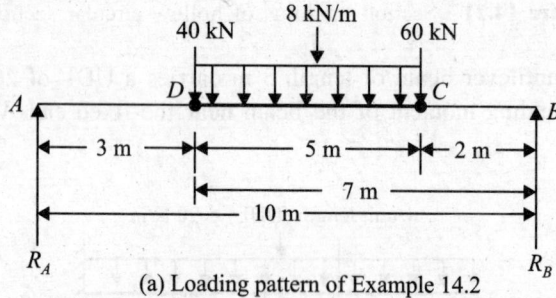

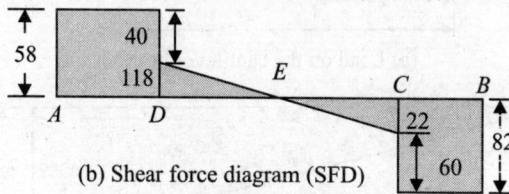

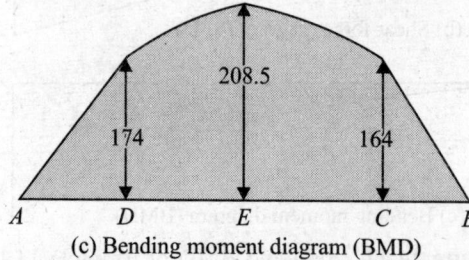

(a) Loading pattern of Example 14.2

(b) Shear force diagram (SFD)

(c) Bending moment diagram (BMD)

**Figure 14.23**   SFD and BMD of Example 14.2.

To draw BMD, (Figure 14.23(c))
BM at $A = 0$. It varies linearly to $D$. At $D$ it is equal to $R_A \times 3 = 58 \times 3 = 174$ kNm.
Again $M$ at $C = 58 \times 8 - 40 \times 5 - 8 \times 5 \times 2.5 = 164$ kNm
At $E$ shear force is zero. Let $E$ be at a distance $x$ from $D$.
Then

$$(58 - 40) - 8x = 0$$
or $$18 = 8x$$
or $$x = 2.25 \text{ m}$$

i.e. shear force is zero at a distance $(3 + 3.36) = 5.25$ m from $A$ where BM is maximum.
$M_E = 58 \times 5.25 - 40 \times 2 - 8 \times 2 \times 1 = 208.5$ kNm BM from $D$ to $C$ is parabolic.

***EXAMPLE 14.3***   A cantilever beam of length 4.6 m is subjected to a UDL of 10 kN/m for a length of 2 m free end. Analyze the beam and draw SFD and BMD.

***Solution***   Figure 14.24(a) shows the loading on the given cantilever of Example 14.3.

**For SFD:**   Consider any section between $A$ and $C$ at a distance $x$ from $A$. The shear force at the section is given by:

$$F_x = wx = 10x$$

At $A$, $x = 0$, hence $F_x = 0$ and at $x = 2$ m, $F_x = 20 \times 2 = 20$ kN. Hence shear force follows a straight line between $A$ and $C$ and with a constant force = 20 kN between $C$ and $B$ as shown in Figure 14.24(b).

**For BMD:**   BM at any section $x$ from $A$ between $A$ and $C = B_x = -(wx) \cdot \dfrac{x}{2} = -10 \cdot x \cdot$

$\dfrac{x}{2} = -5x^2$.

At $A$, $x = 0$, $B_x = 0$ and at $x = 2$, $B_x = -20$ kNm. From $A$ to $C$, BM distribution is parabolic. BM at any section between $C$ and $B$ at a distance $x$ from free end

$M_x = -(10 \times 2) \times (x - 1) = -20x + 20$, i.e. between $C$ and $B$, BM varies linearly.

At $x = 2$ m, $M_x = -20$ kNm and $x = 4.6$, $M_x = -72$ kNm.

With these values at $A$, $C$ and $B$, BMD is drawn in Figure 14.24(c)

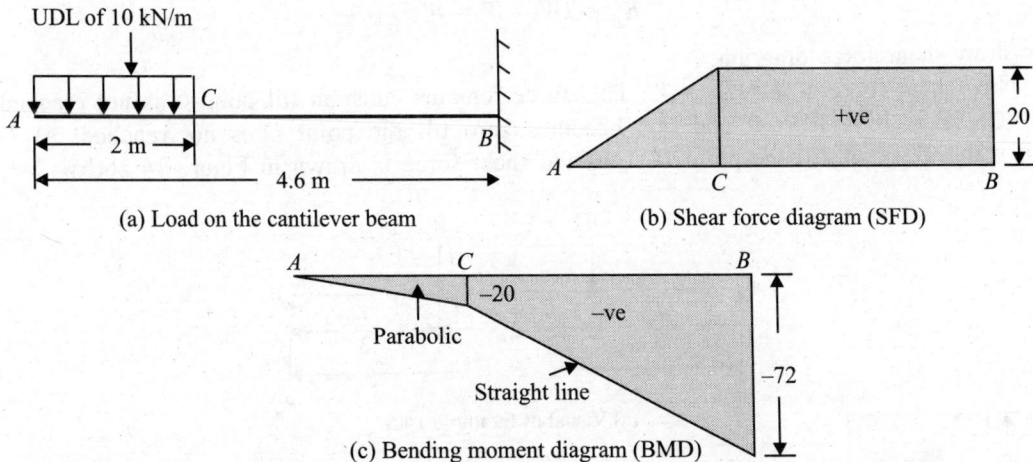

(a) Load on the cantilever beam          (b) Shear force diagram (SFD)

(c) Bending moment diagram (BMD)

**Figure 14.24**   SFD and BMD of Example 14.3.

***EXAMPLE 14.4***   A cantilever beam of length 3 m carries UDL of 40 kN/m throughout the length together with a point load of 40 kN at the free end. Calculate shear force and bending moment on the beam near the fixed end.

***Solution***   Considering Figure 14.25,

Shear force at $x = F_x = 40 + wx = 40 + 40x$

At $x = 3$, i.e. near the fixed end, shear force = $40 + 40 \times 3 = 160$ kN      (Answer)

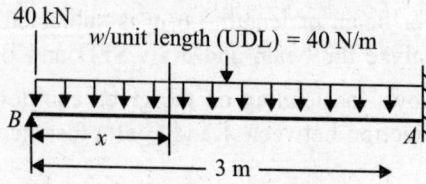

**Figure 14.25**    Visual of Example 14.4.

Bending moment at $x = M_x = -40 \times x - wx\dfrac{x}{2} = -40x - \dfrac{wx^2}{2}$

At $x = 3$, i.e. near the fixed end, bending moment $= -40 \times 3 - 40 \times \dfrac{3^2}{2} = -300 \text{ kNm}$    (Answer)

**EXAMPLE 14.5**    A simply supported beam of length $L$ carries two point load $W$ at $L/3$ from each support. Draw the shear force and bending moment diagram.

**Solution**    The visual of Example 14.5 is shown in Figure 14.26(a).

To calculate the two reactions $R_A$ and $R_B$:

Take moment of all the forces about $B$,

$$R_A \times L = W \times \frac{2L}{3} + W \times \frac{L}{3} = WL$$

$\therefore$ $$R_A = W$$
$\therefore$ $$R_B = 2W - W = W$$

To draw shear force diagram:

Now shear force at $A = R_A = W$. This force remains constant till point $C$ is not reached. At $C$, SF $= W - W = 0$ and this becomes zero till the point $D$ is not reached. At $D$, SF $= 0 - W = -W$ till the point $B$. Thus the shear force is drawn in Figure 14.26(b).

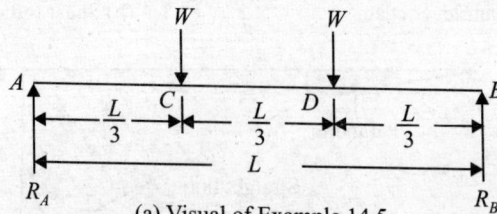

(a) Visual of Example 14.5

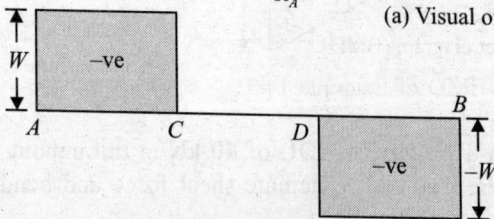

(b) Shear force diagram (SFD)

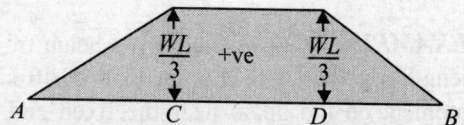

(c) Bending moment diagram (BMD)

**Figure 14.26**    SFD and BMD of Example 14.5.

To draw bending moment diagram:

BM at $A = 0$

BM at $C = M_C = R_A \times \dfrac{L}{3} = \dfrac{WL}{3}$

BM at $D = M_D = R_A \times \dfrac{2L}{3} - W \times \dfrac{L}{3} = \dfrac{2WL}{3} - \dfrac{WL}{3} = \dfrac{WL}{3}$

BM at $B$, $M_B = 0$

With the above values bending moment diagram is drawn in Figure 14.26(c).

**EXAMPLE 14.6**   A beam of 6 m long supported at $A$ and $B$ which are 4 m apart with overhanging of 1 m on each side with two overhanging loads at $C$ and $D$ is shown in Figure 14.27. Draw SF and BM.

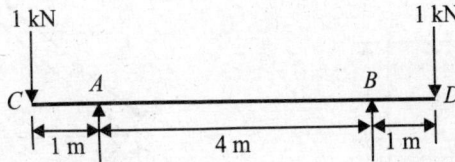

**Figure 14.27**   Overhanging beam of Example 14.6.

**Solution**   From Figure 14.27, reactions $R_A$ and $R_B$ are obtained as:
From the symmetrical loadings and dimensions,

$$R_A = R_B = \frac{1+1}{2} = 1\,\text{kN}$$

Beam with reactions at supports is shown in Figure 14.28(a).
Now shear force at $C = -1$ kN and remains constant between $C$ and $A$.
Shear force between $A$ and $B$ remains constant at zero.
Shear force at $B = 0 + 1$ kN $= 1$ kN and remains constant at 1 kN between $B$ and $D$.
Thus SFD is drawn in Figure 14.28(b).

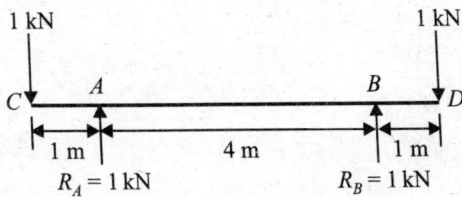

(a) Beam with reactions

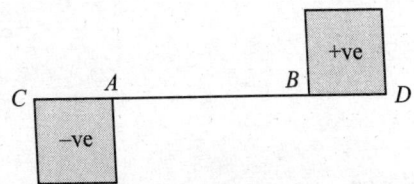

(b) Shear force diagram

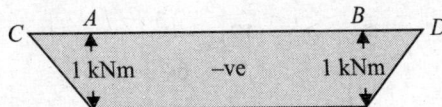

(c) Bending moment diagram

**Figure 14.28**   SFD and BMD of Example 14.6.

BM at $C = 0$

BM at $A = -1 \times 1 = -1$ kNm

BM at any section from $A$ to $B = -1 \times x + R_A \times (x - 1) = -x + x - 1 = -1$ kNm

Hence BM between $A$ and $B$ remains constant at $-1$ kNm.

BM at $D = 0$

Thus BMD is drawn in Figure 14.28(c).

**EXAMPLE 14.7**  For a given stress, compare the moments of resistance of beam of squared section placed with two sides horizontal and with a diagonal-horizontal.

**Solution**  Figures 14.29(a) and 14.29(b) are shown with square section with sides horizontal and diagonal-horizontal respectively.

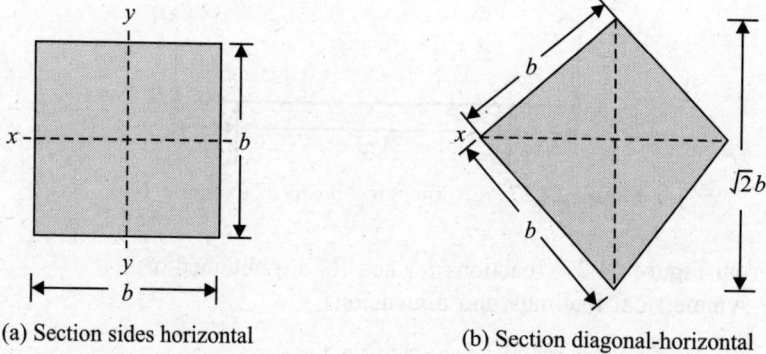

(a) Section sides horizontal           (b) Section diagonal-horizontal

**Figure 14.29**  Visual of Example 14.7.

Let $b$ be the side of the square section.

Section modulus in position (a) of Figure 14.29 = $Z_a \dfrac{1}{6} bd^2 = \dfrac{1}{6} b^3$

In case of (b) of Figure 14.29,

$$I_{xx} = 2 \times \frac{1}{12} b\sqrt{2} \left( \frac{1}{2} b\sqrt{2} \right)^3 = \frac{b^4}{12}$$

$$y = \frac{1}{2} b\sqrt{2} = \frac{b}{\sqrt{2}}$$

Hence

$$Z_b = \frac{\dfrac{b^4}{12}}{\dfrac{b}{\sqrt{2}}} = \frac{b^3}{6\sqrt{2}}$$

Ratio of section moduli gives the comparison of moments of resistance, i.e.

$$\frac{Z_a}{Z_b} = \frac{\frac{1}{6}b^3}{\frac{b^3}{6\sqrt{2}}} = \sqrt{2} = 1.414$$

Hence position of the beam in (a) is 41.4% stronger than position in (b)     (Answer)

## 14.11   CONCLUSION

Different types of beam used in the construction are defined with figures at the beginning. Types of loading on beam like concentrated, uniformly distributed, are explained with illustrative figures. Bending moment and shear force are defined. Sign convention used in the analysis is also explained. Shear force and bending moment diagrams of cantilever, simply supported and overhanging beam are analyzed and drawn. Simple bending theory, bending equation is derived to show the relationship of bending stress, bending moment, moment of inertia, Young's modulus of elasticity and modulus of rigidity. Equations of section modulus of some standard sections are derived. Equation of maximum bending which is dependent on section modulus is also derived. A few illustrative numerical problems are solved and shear force and bending moment diagrams are drawn at the end.

## EXERCISES

**14.1** What are the different types of beam? Explain with figures.

**14.2** What are different types of loading on a beam? Illustrate them with sketches.

**14.3** Define bending moment and shear force and general sign convention used. Explain sagging and hogging bending moment with sketches.

**14.4** Derive bending equations. What are the assumptions?

**14.5** Draw the shear force and bending moment diagrams of a cantilever beam with uniformly loaded throughout the whole span.

**14.6** Find out the equation of maximum bending moment of a simply supported beam when it is subjected to UDL.

**14.7** A simply supported beam of length 8 m carries point loads of 4 kN and 6 kN at distance of 2 m and 4 m from the left end. Draw SFD and BMD of the beam. Also find the maximum BM.                              (Answer   $M_{max}$ = 20 kNm)

**14.8** A cantilever beam of 4 m long carries UDL of 1 kN/m run over the whole length and point load of 2 kN at a distance of 1 m from the free end. Draw SFD and BMD for the beam.                              (Answer   $SF_{max}$ = +14 kN; $BM_{max}$ = –14 kN-m)

# CHAPTER 15

# Hoop Stress: Thin Cylinders and Thin Shells

## 15.1 INTRODUCTION

When a thin cylinder (or shells) is subjected to the action of uniformly distributed radial loading and if the thickness is small compared to the radius $r$, hoop or circumferential distresses are developed. If the loading is directed outwards, the circular ring will be in hoop or circular tension; if it is directed inwards, the ring will be in compression. Such hoop or circumferential stress (tension) acting outwards is quite common in long pipeline carrying water under high internal pressure to power house of hydro-electric project, oil under pressure carrying from one station to another. In the event of quick closer of valve in those pressure pipes, this hoop stress acting outwards becomes very high and thus those pipes must be designed, based on this calculated rise of hoop tensile stress. This distributed stress may occur to internal or external or to centrifugal force as in the case of a rotating ring.

## 15.2 HOOP STRESSES IN THIN CYLINDER

A thin cylindrical shell is shown in Figure 15.1.

Let the internal diameter of the shell = $D$
Thickness of the shell = $t$
Radius of the shell = $r$
Length of the shell = $L$

It is subjected to internal pressure intensity equal to $p$ as shown in Figure 15.1.

Figure 15.2 shows a sectional view of the cylindrical shell shown in Figure 15.1. The axis $x$-$x$ divides the section into two equal parts. Two elementary strips subtending angle $d\theta$ at an angle of $\theta$ on both sides of the vertical through the centre.

Normal force on each strip of length $rd\theta$ is $p \cdot rd\theta \cdot L$ as shown in Figure 15.2. The resultant of these two normal forces acting normal to $x$–$x$ = $p \cdot rd\theta \cdot L \cos \theta + p \cdot rd\theta \cdot L \cos \theta = 2prL \cos d\theta$.

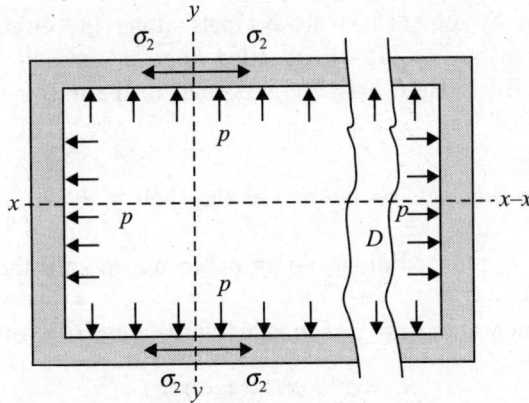

**Figure 15.1** A cylindrical shell subjected to internal pressure intensity $p$.

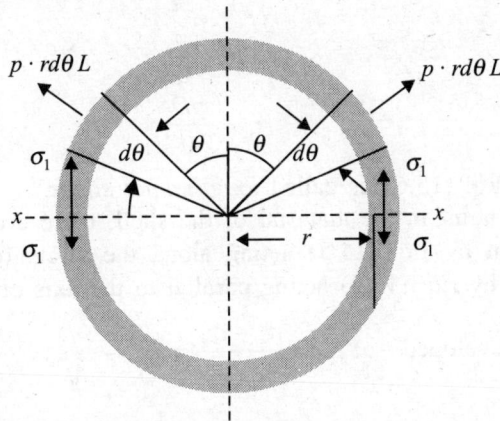

**Figure 15.2** Sectional view of cylindrical shell.

∴ Total force normal to $x$-$x$ on one side of Figure 15.2 is:

$$P = \int_0^{\pi/2} 2prL\cos\theta = 2prL\,|\sin\theta|_0^{\pi/2} = pL(2r)$$

or $\quad\quad\quad P = p\,(LD)$ = pressure × projected area in the upper half $\quad\quad$ (15.1)

Let $\sigma_1$ is the tensile circumferential or Hoop stress induced within the thickness of shell material in the direction of $x$-$x$ as shown in Figure 15.2.

Force of resistance offered by the section $x$-$x = R_x = \sigma_1 \times 2Lt$ $\quad\quad$ (15.2)

Equating Eqs. (15.1) and (15.2),

$$\sigma_1 \times 2Lt = p \cdot LD$$

∴ Hoop or circumferential stress

$$\sigma_1 = \frac{pD}{2t} \quad\quad\quad (15.3)$$

The tensile stress given by Eq. (15.3) always acts along the circumference of the shell. This induced stress, given by Eq. (15.3), is called *hoop or circumferential stress*.

Now consider Figure 15.1. Here axis y-y is normal to the axis x-x. Considering one end (left end) of the section of y-y,

$$\text{Force that acts on the end of the shell} = P = p\,\frac{\pi}{4}D^2 \tag{15.4}$$

A longitudinal stress $\sigma_2$ is induced along y-y in order, not to split the shell in that direction as shown in Figure 15.1.

Thus force of resistance along y-y direction within the material of the shell is:

$$R_y = \sigma_2 \times \pi D \times t = \sigma_2 \pi D t \tag{15.5}$$

Equating (15.4) and (15.5),

$$\sigma_2 \pi D T = p\,\frac{\pi}{4}\,D_2$$

∴ Longitudinal stress

$$\sigma_2 = \frac{pD}{4t} \tag{15.6}$$

The stress $\sigma_2$, given by Eq. (15.6), is called *longitudinal stress*.

It is seen that at any point in the material of the shell, there are two principal stresses. One is hoop stress given by Eq. (15.5) acting along the circumference and the other is longitudinal stress given by Eq. (15.6) acting parallel to the axis of the shell.

$$\text{Maximum shear stress developed} = \tau_{\max} = \frac{\sigma_1 - \sigma_2}{2} = \frac{pD}{2t} - \frac{pD}{4t} = \frac{pD}{8t} \tag{15.7}$$

If $E$ is the Young's modulus and $\dfrac{1}{m} = \mu$ is the Poisson's ratio,

then circumferential strain is $= e_1 = \dfrac{\sigma_1}{E} - \dfrac{\sigma_2}{mE} = \dfrac{pD}{2tE} - \dfrac{pD}{4tEm}$

or circumferential strain $e_1 = \dfrac{pD}{2tE}\left(1 - \dfrac{1}{2m}\right) = \dfrac{pD}{4tE}\left(2 - \dfrac{1}{m}\right)$ $\tag{15.8}$

Longitudinal strain $= e_2 = \dfrac{\sigma_2}{E} - \dfrac{\sigma_1}{mE} = \dfrac{pD}{4tE} - \dfrac{pD}{2tEm} = \dfrac{pD}{2tE}\left(\dfrac{1}{2} - \dfrac{1}{m}\right)$ $\tag{15.9}$

Again circumferentail strain $= e_1 = \dfrac{\text{Change in circumference}}{\text{Original circumference}}$

Circumferential strain $= e_1 = \dfrac{\pi \cdot dD}{\pi \cdot D} = \dfrac{dD}{D}$ $\tag{15.10}$

But $\dfrac{dD}{D}$ = strain in the diameter

$\therefore$  Change in diameter = $e_1 \times$ Original diameter

Longitudinal strain = $\dfrac{\text{Change in length}}{\text{Original length}} = \dfrac{dL}{L} = e_2$

$\therefore$  Change in length = $e_2 \times$ Original length

Volume or capacity of the cylindrical shell = $V = \dfrac{\pi}{4} D^2 L$

Change in volume = $dV = \dfrac{\pi}{4} D^2 dL + \dfrac{\pi}{4} \cdot 2LD dD$

$$\frac{dV}{V} = \frac{\dfrac{\pi}{4} D^2 dL + \dfrac{\pi}{4} 2LD dD}{\dfrac{\pi}{4} D^2 L}$$

Now using Eqs. (15.10) and (15.11),

$$\frac{dV}{V} = \frac{dL}{L} + \frac{dD}{D} = e_2 + 2e_1 \tag{15.11}$$

Again substituting the values of $e_2$ and $e_1$ from Eqs. (15.9) and (15.8) in Eq. (15.11)

$$\frac{dV}{V} = e_2 + 2e_1 = \frac{pD}{2tE}\left(\frac{1}{2} - \frac{1}{m}\right) + 2 \cdot \frac{pD}{4tE}\left(2 - \frac{1}{m}\right)$$

or $$\frac{dV}{V} = \frac{pD}{2tE}\left(\frac{1}{2} - \frac{1}{m} + 2 - \frac{1}{m}\right)$$

or $$\frac{dV}{V} = \frac{pD}{2tE}\left(\frac{5}{2} - \frac{2}{m}\right)$$

$\therefore$ $$dV = \frac{pD}{2tE}\left(5 - \frac{4}{m}\right) \cdot V \tag{15.12}$$

If the fluid flowing through the cylinder is compressible, then bulk modulus of compression $K$ of the fluid to pressure rise (for example water hammer situation) is to be considered in change of volume. Equation (15.12) in such situation becomes:

$$dV = \frac{pD}{2tE}\left(5 - \frac{4}{m}\right) \cdot V + \frac{p}{K} V \tag{15.13}$$

Thus Eqs. (15.12) and (15.13) help to determine in the change the volume or capacity of fluid flowing under radial pressure $p$.

Let $\sigma$ be the permissible tensile stress of the material of the shell. In order that the material should be safe against major principal stress $\sigma_1$, it should be:

$$\frac{pD}{2t} \leq \sigma$$

$$\therefore \qquad t \geq \frac{pD}{2\sigma} \qquad (15.14)$$

Equation (15.14) gives value of thickness for a given radial pressure $p$, diameter and given permissible stress of the material of the cylinder.

## 15.3   STRESSES IN THIN SPHERICAL SHELL

Figure 15.3 shows a thin spherical shell of internal diameter $D$ and thickness $t$. It is subjected to internal pressure intensity of $p$. Now consider the section $x$-$x$ through the centre of the shell $O$. This internal pressure intensity $p$ exerts a bursting force $P_B$ on the wall of the spherical shell.

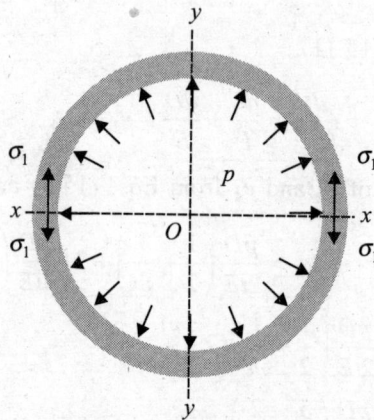

**Figure 15.3**   Thin spherical shell of internal diameter $D$ and thickness $t$ subjected to internal presssure $p$.

Now

$$P_B = p \times \text{projected area} = p \times \frac{\pi}{4} D^2 \qquad (15.15)$$

Let $\sigma_t$ (shown in Figure 15.3) be the tensile hoop stress induced in the shell material in section $x$-$x$. This stress offers the force of resistance $F_R$ to $P_B$.

$$F_R = \sigma_t \cdot \pi D \cdot t \qquad (15.16)$$

Equating (15.15) with (15.16)

$$\sigma_t \cdot \pi D \cdot t = p \cdot \frac{\pi}{4} D^2$$

or hoop stress (tensile)

$$\sigma_t = \frac{pD}{4t} \qquad (15.17)$$

Similarly, it can be shown that tensile stress in any other section through centre $O$ of the shell will have the same value given by Eq. (15.17). This shows that in case of thin spherical shell, principal tensile stresses $\sigma_1 = \sigma_2 = \sigma_t$ and thus no shear stress exists anywhere in the shell. The strain in any section

$$e = \frac{\sigma_t}{E} - \frac{\sigma_t}{mE} = \frac{\sigma_t}{E}\left(1 - \frac{1}{m}\right) = \frac{pD}{4tE}\left(1 - \frac{1}{m}\right) \tag{15.18}$$

Again

$$e = \frac{dD}{D} = \frac{pD}{4tE}\left(1 - \frac{1}{m}\right)$$

and volume of the spherical shell $= V = \dfrac{4}{3}\pi r^3 = \dfrac{4}{3}\pi\left(\dfrac{D}{2}\right)^3 = \dfrac{\pi D^3}{6}$

or

$$dV = \frac{3\pi D^2\, dD}{6}$$

$$\therefore \qquad \frac{dV}{V} = e_V = \left(\frac{3\pi D^2\, dD}{6} \times \frac{6}{\pi D^3}\right) = 3\frac{dD}{D} = 3e = \frac{3pD}{4tE}\left(1 - \frac{1}{m}\right) \tag{15.19}$$

and

$$dV = 3e_V = \frac{3pD}{4tE}\left(1 - \frac{1}{m}\right)V \tag{15.20}$$

When the compressibility of fluid in the sphere is considered, Change in volume will be:

$$dV = 3e_V = \frac{3pD}{4tE}\left(1 - \frac{1}{m}\right)V + \frac{p}{K}V \tag{15.21}$$

When the spherical shell is riveted with efficiency $\eta$ (circumferential) of rivet, stress $\sigma_t$ is given by:

$$\sigma_t = \frac{pD}{2t \cdot \eta_C} \tag{15.22}$$

and for longitudinal stress

$$\sigma_t = \frac{pD}{4t \cdot \eta_L} \tag{15.23}$$

Equations (15.22) and (15.23) have shown that stress is directly proportional to diameter. Stress will be less if the diameter used is less. Therefore, maximum diameter to be used for less internal pressure is the smaller diameter.

**EXAMPLE 15.1**   Water is carried by a cast iron pipe of diameter of 80 cm to the power house under a pressure 200 N/cm². The permissible tensile stress of the material of the pipe is 10000 N/cm². Find the required thickness of the pipe.

**Solution**   Given data are:

$$p = 200 \text{ N/cm}^2$$
$$D = 80 \text{ cm}$$
$$\sigma_1 = 10000 \text{ cm}^2$$

Use Eq. (15.3) to determine the thickness $t$.

$$\sigma_1 = \frac{pD}{2t}$$

or

$$10000 = \frac{200 \times 80}{2t}$$

Solving for thickness $t$,

$$t = 0.8 \text{ cm} \quad \text{(Answer)}$$

**EXAMPLE 15.2** Estimate the safe bursting pressure intensity of a long pipe due to water hammer situation owing to sudden no load condition of the power demand in the powerhouse of high head water power plant, when the pipe diameter is 2.5 m with a thickness of 15 mm. The maximum hoop or circumferential stress and axial stress are 100 N/mm² and 75 N/mm² respectively.

**Solution** Data given in the problem are:

$$D = 2.5 \text{ m} = 2500 \text{ mm}$$
$$t = 15 \text{ mm}$$
$$\sigma_1 = 100 \text{ N/mm}^2$$
$$\sigma_2 = 75 \text{ N/mm}^2$$

Considering $\sigma_1$,

$$\sigma_2 = \frac{pD}{2t}$$

or

$$100 = \frac{p \times 2500}{2 \times 15}$$

Solving for $p$,

$$p = 1.2 \text{ N/mm}^2$$

Considering

$$\sigma_2 = \frac{pD}{4t}$$

or

$$75 = \frac{p \times 2500}{4 \times 15}$$

Solving for $p$,

$$p = 1.8 \text{ N/mm}^2$$

Safe bursting pressure will be the minimum of the two $p$ values = 1.2 N/mm²  (Answer)

**EXAMPLE 15.3** A cylindrical drum of 3.5 m long is made with an internal diameter of 650 mm and thickness of 12 mm. The drum is subjected to an internal pressure of 3.0 N/mm².

If Young's modulus $E = 2 \times 10^5$ N/mm² and Poisson's ratio $\mu = \dfrac{1}{m} = 0.4,$ determine the increase in volume of the drum.

**Solution** Data given are:

$$L = 3.5 \text{ m} = 3500 \text{ mm}, \quad D = 650 \text{ mm}, \quad t = 12 \text{ mm},$$

$$p = 3.0 \text{ N/mm}^2, \qquad \mu = \frac{1}{m} = 0.4, \qquad E = 2 \times 10^5 \text{ N/mm}^2$$

Circumferential strain by Eq. (15.8):

$$e_1 = \frac{pD}{4tE}\left(2 - \frac{1}{m}\right)$$

or

$$e_1 = \frac{3 \times 650}{4 \times 12 \times E}(2 - 0.4) = \frac{65}{E}$$

Longitudinal strain by Eq. (15.9)

$$e_2 = \frac{pD}{2tE}\left(\frac{1}{2} - \frac{1}{m}\right)$$

or

$$e_2 = \frac{3 \times 650}{2 \times 12 \times E}(0.5 - 0.4) = \frac{8.125}{E}$$

Volumetric strain by Eq. (15.11)

$$e_V = \frac{dV}{V} = e_2 + 2e_1 = \frac{8.125}{E} + \frac{2 \times 65}{E} = \frac{138.125}{E}$$

$\therefore$ Increase in volume

$$dV = \left(\frac{138.125}{E}\right)V = \frac{138.125}{2 \times 10^5} \times \left(\frac{\pi}{4} \times 650^2 \times 3500\right)$$

$$dV = 802097.0783 \text{ mm}^3 = 802.097 \text{ cm}^3 \qquad \text{(Answer)}$$

**EXAMPLE 15.4**  A cylindrical shell is 3.5 m long. The diameter and thickness of the shell are 1.2 m and 12 mm respectively. It is subjected to internal pressure of 1.2 N/mm². If Young's modulus $E = 2 \times 10^5$ N/mm² and Poisson's ratio $\mu = \dfrac{1}{m} = 0.4$, determine:

(a) Hoop stress
(b) Longitudinal stress
(c) Maximum shear stress
(d) Change in length
(e) Change in diameter
(f) Change in volume

**Solution**  Data given are:

$$L = 3.5 \text{ m} = 3500 \text{ mm} \qquad D = 1.2 \text{ m} = 1200 \text{ mm} \qquad t = 12 \text{ mm} \qquad p = 1.2 \text{ N/mm}^2$$

$$E = 2 \times 10^5 \text{ N/mm}^2 \qquad \mu = \frac{1}{m} = 0.4,$$

(a) Hoop or circumferential stress

$$\sigma_1 = \frac{pD}{2t} = \frac{1.2 \times 1200}{2 \times 12} = 60 \text{ N/mm}^2 \qquad \text{(Answer)}$$

(b) Longitudinal stress

$$\sigma_2 = \frac{pD}{4t} = \frac{1.2 \times 1200}{4 \times 12} = 30 \text{ N/mm}^2 \qquad \text{(Answer)}$$

(c) Maximum shear stress developed $= \tau_{max} = \dfrac{\sigma_1 - \sigma_2}{2}$ (by Eq. (15.7))

or $\qquad \tau_{max} = \dfrac{\sigma_1 - \sigma_2}{2} = \dfrac{60 - 30}{2} = 15 \text{ N/mm}^2 \qquad$ (Answer)

Longitudinal strain $= e_2 = \dfrac{\sigma_2}{E} - \dfrac{\sigma_1}{mE}$  [by Eq. (15.9)]

or $\qquad e_2 = \dfrac{\sigma_2}{E} - \dfrac{\sigma_1}{mE} = \dfrac{1}{E}\left(\sigma_2 - \sigma_1\dfrac{1}{m}\right) = \dfrac{1}{E}(30 - 60 \times 0.4) = \dfrac{6}{E}$

(d) Change in length $= e_2 \times$ Original length $= \dfrac{6}{E} \times 3500 = \dfrac{6}{2 \times 10^5} \times 3500 = 0.105 \text{ mm}$

(Answer)

Circumferential strain

$$e_2 = \frac{pD}{2tE}\left(1 - \frac{1}{2m}\right) = \frac{pD}{4tE}\left(2 - \frac{1}{m}\right) \text{ [by Eq. (15.8)]}$$

$$e_1 = \frac{pD}{4tE}\left(2 - \frac{1}{m}\right) = \frac{30}{E}(2 - 0.4) = \frac{48}{E}$$

But circumferential strain $= \dfrac{\pi \cdot dD}{\pi \cdot D} = \dfrac{dD}{D}$ $\qquad\qquad$ (by Eq. (15.10))

(e) Change in diameter $e_1 \times D = \dfrac{48}{E} \times D = \dfrac{48}{2 \times 10^5} \times 1200 = 0.288 \text{ mm} \qquad$ (Answer)

Strain in volume $= e_V = \dfrac{dV}{V} = e_2 + 2e_1$ $\qquad\qquad$ (by Eq. (15.11))

Hence

$$\frac{dV}{V} = e_2 + 2e_1 = \frac{6}{E} + 2 \times \frac{48}{E} = \frac{102}{E}$$

$\therefore$ Change in volume $dV = \dfrac{102}{E} \times V = \dfrac{102}{E} \times \dfrac{\pi}{4} \cdot D^2 \cdot L = \dfrac{102}{2 \times 10^5} \times \pi \times 1200^2 \times 3500$

$$= 8075149.757 \text{ mm}^3 \qquad \text{(Answer)}$$

**EXAMPLE 15.5**  A thin spherical sphere is filled with liquid so that the internal pressure becomes 1.6 N/mm². The diameter and thickness of the sphere are 1.6 m and 10 mm respectively.

If Young's modulus $E = 2 \times 10^5$ N/mm² and Poisson's ratio $\mu = \dfrac{1}{m} = 0.4$, determine:

(a) The increase in diameter $\qquad\qquad$ (b) Change in volume of the sphere

**Solution** Data given are:

$$p = 1.6 \text{ N/mm}^2 \qquad D = 1.6 \text{ m} = 1600 \text{ mm} \qquad t = 10 \text{ mm}$$

$$E = 2 \times 10^5 \text{ N/mm}^2 \qquad \mu = \frac{1}{m} = 0.4$$

Using Eq. (15.17)

$$\sigma_t = \frac{pD}{4t} = \frac{1.6 \times 1600}{4 \times 10} = 64 \text{ N/mm}^2$$

Again using Eq. (15.18), strain in any section $e = \dfrac{\sigma_t}{E}\left(1 - \dfrac{1}{m}\right) = \dfrac{64}{E}(1 - 0.4) = \dfrac{38.4}{E}$

or
$$e = \frac{38.4}{2 \times 10^5} = 1.92 \times 10^{-4}$$

∴ Increase in diameter = $e$ × Original diameter = $1.92 \times 10^{-4} \times 1600 = 0.3072$ mm     (Answer)
Change in volume is given by Eq. (15.20), i.e.

$$dV = 3e_V = 3 \times 1.92 \times 10^{-4} \times \left(\frac{\pi}{6} \times 1600^3\right)$$

or
$$dV = 1235324.497 \text{ mm}^3 \qquad \text{(Answer)}$$

**EXAMPLE 15.6**   A spherical thin cell has its thickness equal to 10 mm. The original diameter of the shell is 32 cm. It is filled with water at atmospheric pressure. If an additional 6500 mm³ of water is pumped into the cylinder, find

(a) The pressure inside the shell wall
(b) Hoop stress induced if Young's modulus $E = 2 \times 10^5$ N/mm² and Poisson's ratio

$$\mu = \frac{1}{m} = 0.4.$$

**Solution**   Data given are

$$t = 10 \text{ mm} \qquad d = 32 \text{ cm} = 320 \text{ mm} \qquad E = 2 \times 10^5$$

$$\mu = \frac{1}{m} = 0.4 \qquad \text{Additional water pumped, i.e. } dV = 6500 \text{ mm}^3$$

Let $p$ be the pressure intensity inside the spherical cell wall after pumping an additional amount of 6500 mm³.
We have the Hoop stress Eq. (15.17) inside the spherical shell as:

$$\text{Hoop stress (tensile) } \sigma_t = \frac{pD}{4t} = \frac{p \times 320}{4 \times 10} = 8p \text{ N/mm}^2$$

Using Eq. (15.18), strain in any section $= e = \dfrac{\sigma_t}{E}\left(1 - \dfrac{1}{m}\right)$

According to Eq. (15.19)

$$\frac{dV}{V} = 3e - \frac{3 \times 8p}{2 \times 10^5}(1 - 0.4) - 7.2 - 10^{-5}p$$

or $\qquad dV = (7.2 \times 10^{-5})p \times V$

or $\qquad 6500 = (7.2 - 10^{-5})p - \frac{}{6} - 320^3 - 7.2 - 10^{-5} - 17157284.68$

(a) Solving for $p$ from above, $P - \dfrac{6500}{7.2 - 10^{-5} - 17157284.68} - 5.26 \text{ N/mm}^2$   (Answer)

(b) Hoop stress induced $= {}_t - 8p - 8 - 5.26 - 42.08 \text{ M/mm}^2$   (Answer)

**EXAMPLE 15.7**   A boiler is subjected to an internal pressure of 2.1 N/mm². The thickness of the boiler plate is 28 mm and permissible tensile stress is 125 N/mm². Find the maximum diameter when efficiency of longitudinal joint is 92% and that of circumferential joint is 45%.

**Solution**   Data given are:

$$p = 2.1 \text{ N/mm}^2 \quad t = 28 \text{ mm}, \qquad \text{permissible tensile stress} = 125 \text{ N/mm}^2$$

Efficiency $\eta_L$ of longitudinal joint = 92% = 0.92, and efficiency of circumferential stress = 45% = 0.45

Equation (15.23) for longitudinal efficiency is

$$= \sigma_t = \frac{pD}{4t \cdot \eta_L}$$

or $\qquad 125 = \dfrac{2.1 \times D}{4 \quad 28. \quad 0.92}$

$\therefore \qquad D = 2133.33 \text{ mm} = 213.333 \text{ cm}$

Equation (15.22) for circumferential efficiency is

$$= \sigma_t = \frac{pD}{2t \cdot \eta_C}$$

or $\qquad 125 = \dfrac{2.1 \times D}{2 \quad 28. \quad 0.45}$

$\therefore \qquad D = 1500 \text{ mm} = 150 \text{ cm}$

Maximum diameter will be the smaller one for less internal pressure.
Hence, maximum diameter $D = 150$ cm   (Answer)

## 15.4   CONCLUSION

Equations of tensile hoop stress and longitudinal tensile stress in the material of thin cylindrical shell due to internal pressure have been derived in a very simple and easy-to-follow step by step method. Equations of longitudinal and circumference strains developed due to this internal pressure intensity are also deduced. These strains so developed allow determining the increase of diameter and volume and equations of this increase of diameter and volume are presented. If the permissible tensile stress of the materials of construction of the shell is known, design thickness of the shell can be determined. Similar analysis in the case of spherical thin shell has also been analyzed. A few numerical problems are solved in order to show the use of the theory presented and some more chapter-end exercises with answers are given at the end of the chapter.

## EXERCISES

**15.1** A thin pipe of diameter 1200 mm is subjected to internal water pressure of 1.8 N/mm$^2$. The permissible stress in the materials of construction is 120 N/mm$^2$. Find the minimum thickness. (Answer  9 mm)

**15.2** Water is carried by a cast iron pipe of diameter of 50 cm to the powerhouse under a pressure 150 N/cm$^2$. The permissible tensile stress of the material of the pipe is 5000 N/cm$^2$. Find the required thickness of the pipe $p = 200$ N/cm$^2$. (Answer  0.75 mm)

**15.3** A cylindrical drum of 3.0 m long is made with an internal diameter of 550 mm and thickness of 10 mm. The drum is subjected to an internal pressure of 2.4 N/mm$^2$. If Young's modulus $E = 2 \times 10^5$ N/mm$^2$ and Poisson's ratio $\mu = \dfrac{1}{m} = 0.4$, determine the increase in volume of the drum. (Answer  32.942 cm$^3$)

**15.4** Estimate the safe bursting pressure intensity of a long pipe due to water hammer situation owing to sudden no load condition of the power demand in the powerhouse of high head water power plant, when the pipe diameter is 2.0 m with a thickness of 10 mm. The maximum hoop or circumferential stress and axial stress are 80 N/mm$^2$ and 60 N/mm$^2$, respectively. (Answer  1.2 N/mm$^2$ )

**15.5** A cylindrical shell is 3.0 m long. The diameter and thickness of the shell are 1000 mm and 10 mm respectively. It is subjected to internal pressure of 1.0 N/mm$^2$. If Young's modulus $E = 2 \times 10^5$ N/mm$^2$ and Poisson's ratio $\mu = \dfrac{1}{m} = 0.4$ determine:

(a) Hoop stress
(b) Longitudinal stress
(c) Maximum shear stress
(d) Change in length
(e) Change in diameter
(f) Change in volume

(Answer (a) 50 N/mm$^2$
(b) 25 N/mm$^2$
(c) 12.5 N/mm$^2$
(d) 0.075 mm
(e) 0.2 mm
(f) 4005530.633 mm$^3$)

**15.6** A thin spherical sphere is filled with fluid so that the internal pressure becomes 1.2 N/mm². The diameter of the sphere is 1500 mm. If the thickness of the spherical shell is 10 mm and Young's modulus $E = 2 \times 10^5$ N/mm² and Poisson's ratio $\mu = \dfrac{1}{m} = 0.4$, determine:

(a) The increase in diameter

(b) Change in volume of the sphere

(Answer  (a)  0.2025 mm

(b)  715694.0764 mm³)

**15.7** The thickness of a boiler plate is 25 mm. It is subjected to an internal pressure of 3.0 N/mm². The permissible tensile stress is 125 N/mm². If the efficiency of longitudinal joint is 90% and that of circumferential joint is 35%, find the maximum diameter.

(Answer  1458.3 mm)

# CHAPTER 16

# Torsion

## 16.1 INTRODUCTION

In Chapter 11, torque has been defined in Section 11.3. It is the turning moment of force on a body. When equal and opposite torques are applied at the two ends of shaft, the shaft is said to be in torsion. The shaft undergoes a twisting moment due to the application of the torques which produces the shear stresses and strains in the material of the shaft. Torsion occurs in machine elements, beams, slabs, shear walls, aircraft structure, ships, shaft transmitting power, springs, box girders, building frames, etc. Therefore, preliminary knowledge of torsion is essential for all students of all disciplines.

## 16.2 ASSUMPTIONS IN TORSIONAL ANALYSIS

The torsional analysis in the derivation of shear stress produced in a circular shaft subjected to torsion is based on the following assumptions. The assumptions are more or less similar to those of bending theory discussed in Chapter 14.

  (i) The material of the shaft is uniform throughout.
 (ii) Sections plane before twisting remain plane after twisting.
(iii) The material is an isotropic homogenous elastic continuum.
(iv) Strains and deformations are small.
 (v) All radii which are straight before twist remain straight after twist.

## 16.3 SHEAR STRESS EQUATION IN A CIRCULAR SHAFT SUBJECTED TO TORSION

A circular fixed shaft which is subjected to torque $T$ at the free end is shown in Figure 16.1. Due to this applied torque $T$ at the free end as shown in the figure, the shaft at this free end rotates. The line which was originally in position $AD$ will be deflected to $AD'$. The distortion of outer surface of the shaft = $DD'$. Thus angle $\theta$ in sectional view is called *angle of twist* and angle $\phi$ in front view is equal to shear strain.

Let $L$ be the length and $R$ be the radius of the shaft. $L$ and $R$ are shown in Figure 16.1(a) i.e. in front elevation and in 16.1(b), i.e. in sectional view.

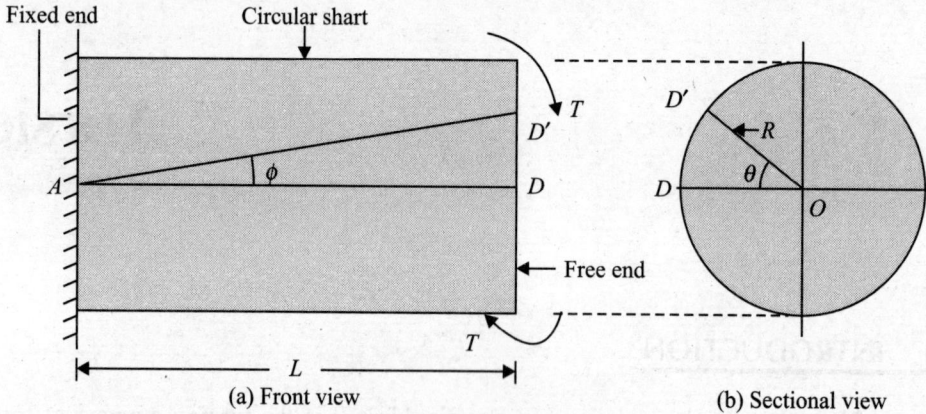

(a) Front view　　　　　　　　　　　(b) Sectional view

**Figure 16.1** Circular fixed shaft subjected to torque $T$ at the free end.

Shear stress $\tau$ is induced on the surface of the shaft as a result of application of torque. Let $G$ be the modulus of rigidity of the material of the shaft.

Now shear strain at outer surface = Distortion per unit length

$$= \frac{DD'}{L} = \tan \phi = \phi \quad \because \phi \text{ is very small.}$$

Hence shear strain at outer surface,

$$= \phi = \frac{DD'}{L} \tag{16.1}$$

From Figure 16.1(b),

$$DD' = OD \times \theta = R\theta \tag{16.2}$$

Substituting the value of $DD'$ from (16.2) in (16.1),
shear strain at outer surface

$$= \phi = \frac{R\theta}{L} \tag{16.3}$$

Now, modulus of rigidity $G$ of the material of the shaft is:

$$G = \frac{\text{Shear stress induced}}{\text{Shear strain produced}}$$

or　　　　　$$G = \frac{\text{Shear stress at outer surface}}{\text{Shear strain at outer surface}}$$

or
$$G = \frac{\tau}{\left(\dfrac{R\theta}{L}\right)} = \frac{\tau \cdot L}{R\theta}$$

or
$$\frac{G\theta}{L} = \frac{\tau}{R} \qquad (16.4)$$

$\therefore$
$$\tau = \frac{RG\theta}{L} \qquad (16.5)$$

For given shaft subjected to given torque $T$, $G$, $L$ and $\theta$ are constant. Hence, shear stress is proportional to radius

$$\tau \, \alpha \, R \qquad (16.6)$$

or
$$\frac{\tau}{R} = \text{constant}$$

Let $\tau_r$ be the shear stress induced at a radius $r$ from the centre of the shaft.
Then
$$\frac{\tau}{R} = \frac{\tau_r}{r} \qquad (16.7)$$

Combining Eq. (16.4) with Eq. (16.7),

$$\frac{\tau}{R} = \frac{G\theta}{L} = \frac{\tau_r}{r} \qquad (16.8)$$

From Eq. (16.6), shear stress is directly proportional to radius. It gives maximum shear stress at surface and zero shear stress at centre of the shaft.

## 16.4  MAXIMUM TORQUE TRANSMITTED BY A SOLID CIRCULAR SHAFT

A sectional view of the shaft of radius $R$ is shown in Figure 16.2. It is already explained that induced shear stress $\tau$ is maximum at the surface, i.e. when radius is $R$. In Figure 16.3, an elementary circular ring of thickness $dr$ at a distance $r$ from centre is considered. Shear stress at radius $r$ is taken as $\tau_r$.

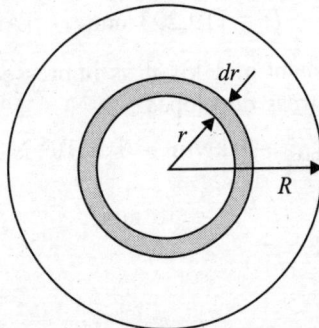

**Figure 16.2**  Sectional view of the shaft.

Area of this ring is

$$dA = 2\pi \cdot r \cdot dr$$

Shear stress at $r$ from Eq. (16.7)

$$\tau_r = \frac{\tau}{R} r$$

Turning force on this small ring $= \tau_r \times dA = \tau_r \times 2\pi \cdot r \cdot dr = \frac{\tau}{R} r \times 2 \cdot \pi \cdot r \cdot dr = 2\pi \frac{\tau}{R} r^2 dr$

Turning moment on this small ring $=$ Turning force $\times$ Distance of ring from centre

$$= 2\pi \frac{\tau}{R} r^2 dr \times r = 2\pi \frac{\tau}{R} r^3 dr \tag{16.9}$$

Total turning moment or total or maximum torque is obtained by integrating term in Eq. (16.1) from zero to maximum radius.

or

$$T_{\max} = \int_0^R 2\pi \frac{\tau}{R} r^3 dr = 2\pi \frac{\tau}{R} \left[ \frac{r^4}{4} \right]_0^R = \frac{\pi \cdot \tau \cdot R^3}{2} = \frac{\pi \cdot \tau}{2} \left( \frac{D}{2} \right)^3 = \frac{\pi}{16} \tau \cdot D^3 \tag{16.10}$$

**EXAMPLE 16.1**   Determine the diameter of solid shaft which has to transmit a torque of 20 kN-m if the maximum permissible shear stress is 60 N/mm²

**Solution**   Maximum shear stress on the surface of the shaft $= \tau = 60$ N/mm²
Torque to transmit $= T_{\max} = 20$ kN-m $= 20 \times 10^6$ N-mm
Let $D$ be the diameter of the shaft in mm
Using Eq. (16.10),

$$T_{\max} = \frac{\pi}{16} \tau \cdot D^3$$

or

$$20 \times 10^6 = \frac{\pi}{16} \times 60 \times D^3$$

or

$$D^3 = \frac{16 \times 20 \times 10^6}{\pi \times 60}$$

$\therefore$

$$D = 119,293 \text{ mm} \qquad \text{(Answer)}$$

**EXAMPLE 16.2**   A twisting moment of 1 kN-m is impressed upon a 50 mm diameter shaft. What is the maximum shearing stress developed?

**Solution**   Twisting moment or $T_{\max} = 1$ kN-m $= 1 \times 10^6$ N-mm
Diameter

$$D = 50 \text{ mm}$$

Let the shearing stress developed $= \tau$
Using Eq. (16.10),

$$T_{\max} = \frac{\pi}{16} \tau \cdot D^3$$

or
$$1 \times 10^6 = \frac{\pi}{16} \tau \times 50^3$$

∴
$$\tau = 40.743 \text{ N/mm}^2 \quad \text{(Answer)}$$

## 16.5 TORQUE TRANSMITTED BY HOLLOW CIRCULAR SHAFT

A sectional view of a hollow shaft of radius $R_o$ and $R_i$ is shown in Figure 16.3. As shown in figure $R_o$ and $R_i$ are the outer and inner diameters of the shaft respectively.

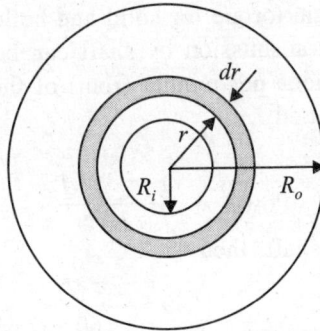

**Figure 16.3** Sectional view of a hollow shaft.

Considering the elemental ring at a distance $r$ from the centre, Eq. (16.9) in Section 16.4 will be in terms $R_o$.

Turning force on this small ring = $\tau_r \times dA = \tau_r \times 2\pi \cdot r \cdot dr = \frac{\tau}{R_o} r \times 2 \cdot \pi \cdot r \cdot dr = 2\pi \frac{\tau}{R_o} r^2 dr$.

Integrating the term $2\pi \dfrac{\tau}{R_o} r^2 dr$ within the limit of $R_i$ and $R_o$.

Final equation of $T_{max}$ in terms outside and inside diameters $D_o$ will $D_i$ be obtained after simplification as:

$$T_{max} = \frac{\pi}{16} \tau \left( \frac{D_o{}^4 - D_i{}^4}{D_o} \right) \tag{16.11}$$

***EXAMPLE 16.3*** Calculate the maximum shear stress developed in a hollow shaft if the torque transmitted through the shaft is 5 kN-m. The inside and outside diameters are 60 mm and 100 mm respectively.

***Solution*** For maximum shear stress, torque developed = $T_{max}$ = 5 kN-m = $5 \times 10^6$ N-mm

$$\text{Outside diameter } D_o = 100 \text{ mm}$$
$$\text{Inside diameter } D_i = 60 \text{ mm}$$

Using Eq. (16.11),

$$T_{max} = \frac{\pi}{16} \tau \left( \frac{D_o{}^4 - D_i{}^4}{D_o} \right)$$

or
$$5 \times 10^6 = \frac{\pi}{16} \tau \left( \frac{100^4 - 60^4}{100} \right).$$

or
$$\frac{5 \times 10^6 \times 16}{\pi} = 870400 \times \tau$$

$$\tau = 29.256 \text{ N/mm}^2 \qquad \text{(Answer)}$$

## 16.6  POWER TRANSMITTED BY SHAFT

From the expressions for maximum torque for solid and hollow shafts given by Eqs. (16.10) and (16.11), equation of power transmission by shaft can be obtained.

Let $N$ be the speed in revolutions per minute (rpm) of the shaft.

$T$ be the mean torque transmitted

Then

$$\text{Power} = \frac{2 \cdot \pi \cdot N \cdot T}{60} \tag{16.12}$$

If $\omega$ is the angular speed of the shaft, then

$$\omega = \frac{2\pi \cdot N}{60}$$

Hence

$$\text{Power} = \omega \cdot T \tag{16.13}$$

If mean torque $T$ is in N-m, unit of power will be in watt.

**EXAMPLE 16.4**  A solid circular shaft of steel is 50 mm in diameter. Find the power transmitted at 120 rpm if the permissible shear stress is 62.5 N/mm$^2$.

**Solution**  Permissible shear stress given = $\tau = 62.5$ N/mm$^2$
Diameter of the shaft = 50 mm
Speed of shaft = 120 rpm
Using Eq. (16.10), permissible torque can be evaluated

or
$$T_{\max} = \frac{\pi}{16} \tau \cdot D^3 = \frac{\pi}{16} \times 62.5 \times 50^3 = 1533980.788 \text{ N-mm}$$

or
$$T_{\max} = 1533980.788 \text{ N-mm} = 1.53398 \text{ N-m}$$

Using Eq. (16.12) for power transmitted = $\dfrac{2 \cdot \pi \cdot N \cdot T}{60}$

or
$$\text{Power} = \frac{2\pi \times 120 \times 1.53398}{60} = 19.276 \text{ Watt} \qquad \text{(Answer)}$$

**EXAMPLE 16.5**  A solid shaft has transmitted 75 kW at 200 rpm. If the allowable shear stress is 70 N/mm$^2$, find the suitable diameter for the shaft, if the maximum torque transmitted at each revolution exceeds the mean by 30%.

*Solution*   Data given are:

Power transmitted = 75 kW = 75000 W, N = 200 rpm, shear stress $\tau = 70 \text{N/mm}^2$

If $T$ is the average torque, $T_{max} = 1.3$ T

Let $D$ be the suitable diameter.

The power Eq. (16.12) is:

$$\text{Power} = \frac{2 \cdot \pi \cdot N \cdot T}{60}$$

or
$$75000 = \frac{2\pi \times 200 \times T}{60}$$

∴
$$T = 3580.98622 \text{ N-m} = 3580986.22 \text{ N-mm}$$
$$T_{max} = 1.3 \text{ T} = 1.3 \times 3580986.22 = 4655282.086$$

Using Eq. (16.10),

$$T_{max} = \frac{\pi}{16} \tau \cdot D^3$$

$$4655282.086 = \frac{\pi}{16} \times 70 \times D^3$$

∴
$$D = 69.7063 \text{ mm} \qquad \text{(Answer)}$$

**EXAMPLE 16.6**   A hollow shaft has to transmit 300 kW power at 80 rpm. If the shear stress is not to exceed 60 N/mm² and internal diameter is 60% of the external diameter, find the external and internal diameters when maximum torque is 1.4 times the average torque and thickness of the shaft.

*Solution*   Data given are: Power = 300 kW = 300000 W, speed of the shaft $N = 80$ rpm, maximum shear stress $\tau = 60$ N/mm², Internal diameter = $D_i = 0.6\ D_o$, $T_{max} = 1.4$ T

Using the power Eq. (16.12):

$$\text{Power} = \frac{2 \cdot \pi \cdot N \cdot T}{60}$$

or
$$300000 = \frac{2 \times \pi \times 80 \times T}{60}$$

∴
$$T = 35809.8622 \text{ N-m}$$

Hence

$$T_{max} = 1.4 \times 35809.8622 = 50133.807 \text{ N-m} = 50133807 \text{ M-mm}$$

Using Eq. (16.11),

$$T_{max} = \frac{\pi}{16} \tau \left( \frac{D_o^{\,4} - D_i^{\,4}}{D_o} \right)$$

or
$$50133807 = \frac{\pi \times 60 \times [D_o^{\,4} - (0.6 D_{oh})^4]}{16 \times D_o}$$

or
$$50133807 \times 16 = 188.4955 \times 0.8704 \times D_o^{\,3}$$

∴
$$D_o = 169.724 \text{ mm} \qquad \text{(Answer)}$$

and $D_i = 0.6\,D_o = 0.6 \times 169.724 = 101.834$ mm    (Answer)

Thickness of the shaft $= \dfrac{D_o - D_i}{2} = \dfrac{169.724 - 101.834}{2} = 33.9415$ mm    (Answer)

***EXAMPLE 16.7***  A solid shaft transmits 90 kW at 180 rpm. Modulus of rigidity of the shaft is $8 \times 10^4$ N/mm$^2$. The maximum shear stress is limited to 60 N/mm$^2$. Determine the diameter of the shaft. Also determine its length if the twist must not exceed 1° over the entire length.

***Solution***  Data given are: Power $= 90$ kW $= 90000$ W, $N = 160$ rpm, shear stress $\tau = 60$ N/mm$^2$, modulus of rigidity $G = 8 \times 10^4$ N/mm$^2$, angle of twist $= \theta = 1°$ or $\dfrac{\theta}{180}$ radian.

Let $D$ and $L$ be the diameter and length of the shaft respectively.
Using the power Eq. (16.12):

$$\text{Power} = \frac{2\pi \cdot N \cdot T}{60}$$

or $$90000 = \frac{2 \times \pi \times 160 \times T}{60}$$

$\therefore$ $$T = 5371.479 \text{ N-m} = 5371479 \text{ N-mm}$$

Assuming this torque to be $T_{max}$ and using Eq. (16.10),

$$T_{max} = \frac{\pi}{16}\,\tau \cdot D^3$$

or $$5371479 = \frac{\pi}{16} \times 60 \times D^3$$

$\therefore$ $$D = 76.967 \text{ mm} \quad \text{(Answer)}$$

Using Eq. (16.8)

$$\frac{\tau}{R} = \frac{G\theta}{L}$$

or $$\frac{60}{\dfrac{76.967}{2}} = \frac{8 \times 10^4 \times \pi}{L \times 180} \qquad \because \theta = \frac{\pi}{180} \text{ radian}$$

Solving for $L$,

$$L = 895.55 \text{ mm} \quad \text{(Answer)}$$

## 16.7  POLAR MOMENT OF INERTIA

Polar moment of inertia of a plane area is defined as the moment of inertia of the area about an axis perpendicular to the plane of the figure and passing through the centre of gravity of the area. Polar moment of inertia is normally denoted by $J$.

This $J$ can be related to torque $T$ using Eq. (16.9), i.e.

Turning moment on this small ring $= dT = 2\pi \dfrac{\tau}{R} r^3 dr$

or
$$dT = \frac{\tau}{R} r^2 (2\pi \cdot r dr)$$

or
$$dT = \frac{\tau}{R} r^2 dA$$

or
$$T = \int_0^R dT = \frac{\tau}{R} \int_0^R r^2 dA \tag{16.14}$$

The term $r^2 dA$ is the moment of inertia of the elementary ring about an axis perpendicular to the plane of Figure 16.2 and passing through the centre of the circle.

Hence $\displaystyle\int_0^R r^2 dA$ is the moment of inertia of the whole circular area about an axis perpendicular to the plane of Figure 16.2 and passing through the centre of the circle.

Thus $\displaystyle\int_0^R r^2 dA$ = Polar moment of area $J = \dfrac{\pi}{32} D^4$

$\therefore$ From Eq. (16.14)

$$T = \frac{\tau}{R} J \tag{16.15}$$

or
$$\frac{T}{J} = \frac{\tau}{R} \tag{16.16}$$

A relationship among $J$, $T$, $R$, $L$, $G$, $\theta$ and $\tau$ may be obtained as:
Equation (16.4) is:

$$\frac{G\theta}{L} = \frac{\tau}{R}$$

Combining Eqs. (16.4) and (16.16),

$$\frac{\tau}{R} = \frac{G\theta}{L} = \frac{T}{J} \tag{16.17}$$

## 16.8 STRENGTH OF SHAFT AND TORSIONAL RIGIDITY

The maximum torque or maximum power that can be transmitted is called the *strength of the shaft*. The product of modulus of rigidity $G$ and polar moment of inertia $J$ is called *torsional rigidity* or *stiffness of the shaft*. Thus mathematically,

Torsional rigidity = $GJ$ (16.18)

Torsional rigidity is also defined as the torque required producing a twist of one radian per unit length of the shaft. This definition can be proved as:

Let a twisting moment or torque $T$ produce a twist of radians in a shaft of length $L$. Using Eq. (16.17),

$$\frac{\tau}{R} = \frac{G\theta}{L} = \frac{T}{J}$$

or

$$G \cdot J = \frac{TL}{\theta}$$

By Eq. (16.18),

$$G \cdot J = \text{Torsional rigidity}$$

Hence

$$\text{Torsional rigidity} = \frac{TL}{\theta} \tag{16.19}$$

In Eq. (16.19), if $L = 1$ m, $\theta = 1$ radian

$$\text{Torsional rigidity} = \text{Torque } T$$

**EXAMPLE 16.8**   A solid shaft transmits 300 kW at 250 rpm. The length of the shaft is 2 m. The maximum shear stress must not exceed 60 N/mm², the modulus of rigidity is $1 \times 10^5$ N/mm². Determine the diameter of the shaft if angle of twist must not exceed 1° over the entire length.

**Solution**   Data given are:

Power = 300 kW = 300000 W, shaft speed $N$ = 250 rpm, maximum shear stress = $\tau$ = 30 N/mm$_2$, length of the shaft = 2 m = 2000 mm, modulus of rigidity $G = 1 \times 10^5$ N/mm$_2$ Twist

of the shaft $\theta = 1° = \dfrac{\pi}{180} = 0.01745329$ radian.

Let $D$ be the diameter which is to be determined from the above data.
Using the power Eq. (16.12):

$$\text{Power} = \frac{2 \cdot \pi \cdot N \cdot T}{60}$$

or

$$300000 = \frac{2 \cdot \pi \times 250 \times T}{60}$$

∴              $T = 11459.1559$ M-m $= 11459155.9$ m

Diameter $D$ can be worked out from shear stress and from angle of twist $\theta$.
Diameter $D$ with shear stress:
Using the equation

$$T = \frac{\pi}{16} \tau \cdot D^3$$

or

$$11459155.9 = \frac{\pi}{16} \times 30 \times D^3$$

∴              $D = 124.834$ mm

Diameter from angle of twist:
Using the equation

$$\frac{G\theta}{L} = \frac{T}{J}$$

or
$$\frac{1 \times 10^5 \times 0.01745329}{2000} = \frac{11459155.9}{\frac{\pi}{32} D^4}$$

or
$$0.085673636 \, D^4 = 11459155.9$$

∴
$$D = 107.541 \text{ mm}$$

Out of these two diameters, greater value is to be taken. If smaller is selected, then by equation, $T = \frac{\pi}{16} \tau \cdot D^3$, shear stress value goes above the given value 30 N/mm², as $D$ is smaller.

Hence correct diameter will be

$$D = 124.834 \text{ mm} \qquad \text{(Answer)}$$

## 16.9 POLAR MODULUS

Polar modulus or torsional section modulus is defined as polar moment of inertia to the radius of the shaft. If $Z_p$ is the polar modulus, mathematically,

$$Z_p = \frac{J}{R}$$

or
$$Z_p = \frac{\frac{\pi}{32} D^4}{\frac{D}{2}} = \frac{\pi}{16} D^3 \qquad (16.20)$$

Similarly it can be derived for hollow shaft as:

$$Z_p = \frac{\pi}{16 D_o} (D_o^4 - D_i^4) \qquad (16.21)$$

## 16.10 COMBINED BENDING AND TORSION

A shaft is subjected to shearing stresses in the process of transmitting torque or power. At the same time, the shaft is also subjected to bending moments due to gravity or inertia forces. This bending moment produces bending stresses in the shaft. Therefore, each particle of the shaft is subjected to both shear and bending stresses.

Shear stress $\tau_r$ at any point due to torque $T$ is given by Eq. (16.16),

$$\frac{T}{J} = \frac{\tau}{R} = \frac{\tau_r}{r}$$

or

$$\tau_r = \frac{T}{J}r \qquad (16.22)$$

Equation of bending stress $f$ due to moment $M$ is given by Eq. (14.22) in Chapter 14 as:

$$\frac{M}{I} = \frac{f}{y} = \frac{E}{R}$$

or

$$f = \frac{M}{I}y \qquad (16.23)$$

Shear and bending stresses are the maximum on the surface, i.e. $r = \dfrac{D}{2}$.
Hence from Eq. (16.22),

$$\tau_{max} = \frac{T}{J} \times \frac{D}{2} = \frac{T}{\dfrac{\pi}{32}D^4} \times \frac{D}{2} = \frac{16T}{\pi D^3} \qquad (16.24)$$

And from Eq. (16.23),

$$f_{max} = \frac{M}{I} \times \frac{D}{2} = \frac{M}{\dfrac{\pi}{64}D^4} \times \frac{D}{2} = \frac{32M}{\pi D^3} \qquad (16.25)$$

Again we have the relationship of direct stress $\sigma$ or $f$ shear stress $\tau$ and angle $\theta$ made by the plane of maximum shear with the normal cross section given by,

$$\tan 2\theta = \frac{2\tau}{f} \qquad \text{(Eq. (13.20) when only one stress)}$$

When shear and bending stress is the maximum,

$$\tan 2\theta = \frac{2\dfrac{16T}{\pi D^3}}{\dfrac{32M}{\pi D^3}} = \frac{T}{M} \qquad (16.26)$$

Major principal stress $= \dfrac{f_{max}}{2} + \sqrt{\left(\dfrac{f_{max}}{2}\right)^2 + \tau^2}$ \qquad (Eq. (13.18) when one stress)

$$= \frac{32M}{2 \times \pi D^3} + \sqrt{\left(\frac{32M}{2 \times \pi D^3}\right)^2 + \left(\frac{16T}{\pi D^3}\right)^2}$$

$$= \frac{16}{\pi D^3}(M + \sqrt{M^2 + T^2}) \qquad (16.27)$$

Similarly minor principal stress $= \dfrac{f_{max}}{2} - \sqrt{\left(\dfrac{f_{max}}{2}\right)^2 + \tau^2}$

$$= \frac{16}{\pi D^3}\left(M - \sqrt{M^2 + T^2}\right) \tag{16.28}$$

## 16.11  CONCLUSION

Definition of torsion and its application in different fields of engineering is mentioned in the beginning. Assumptions made in the analysis of torsion are made. The important equation of shear stress analysis in torsion is derived. Equation of maximum torque relating shear stress and diameter of the solid shaft is also deduced. Hollow shaft is considered and corresponding equation is also given. Power transmitted by shaft, polar moment of inertia, polar modulus, strength of shaft and combined bending and shearing strength are discussed with related equation. A few numerical examples are solved to show the application of mathematical equation derived in the chapter. Also some chapter-end exercises are given for student to practice.

## EXERCISES

**16.1** Derive an expression for the shear stress produced in a circular shaft which is subjected to torsion. What are the assumptions made in the derivation?

**16.2** Derive an expression for angle of twist of a circular shaft as a function of applied twisting moment.

**16.3** Derive an expression for a circular shaft in the following form with usual notations:

$$\frac{\tau}{R} = \frac{G\theta}{L} = \frac{T}{J}$$

**16.4** Deduce the expression for the torque transmitted by a hollow circular shaft in terms of its external and internal diameters.

**16.5** Define polar modulus. Find the expressions of polar modulus for solid and hollow shafts.

**16.6** A solid shaft of 200 mm is used to transmit torque. If the maximum shear stress induced in the shaft is 50 N/mm², find maximum torque transmitted.

(Answer   78.539.816 N-m)

**16.7** The shearing stress in a solid shaft is not to exceed 50 N/mm² when the torque transmitted is 45000 N-m. Determine the minimum diameter of the shaft.

(Answer   166.113 mm)

**16.8** A hollow shaft has external and internal diameters 320 mm and 160 mm respectively. If the shear stress is not to exceed 42 N/mm², find the maximum torque transmitted by this hollow shaft.    (Answer   25333803.16 N-mm = 253.338 kN-m)

# CHAPTER 17

# Columns and Struts

## 17.1 INTRODUCTION

Columns and struts are the members of a structure. Both these members are subjected to axial compressive load. Columns are vertical members and its both ends are fixed or one end fixed other end hinged or one end fixed other end free or both ends hinged. A simple example of column is a vertical pillar of a building between the roof and the floor in which the ends are fixed. If the members are not vertical and ends are hinged or pin-jointed, they are called *struts*. Examples of struts are piston rod, connecting rods, etc. Columns are not usually short. Their behaviour is governed more by stability than strength. In this chapter, attempts will be made to present its failure by crushing or buckling due to load, behaviour of column and struts under different end conditions, mathematical and empirical equations of design, their buckling concept, theory of crippling load for different end conditions, etc.

## 17.2 FAILURE OF COLUMN

Columns are load-bearing members. Therefore, different stresses are developed in columns. These stresses are main reasons of failure of columns. They may fail due to any of the following stresses.

   (i) Direct compressive stresses
   (ii) Buckling stresses
  (iii) Combination of direct and buckling stresses.

## 17.3 FAILURE OF SHORT AND LONG COLUMNS

A column is called *short column* if its length in comparison to cross-sectional area is small as shown in Figure 17.1(a). A column is called *long column* if its length in comparison to its lateral dimensions is very large as shown in Figure 17.1(b).

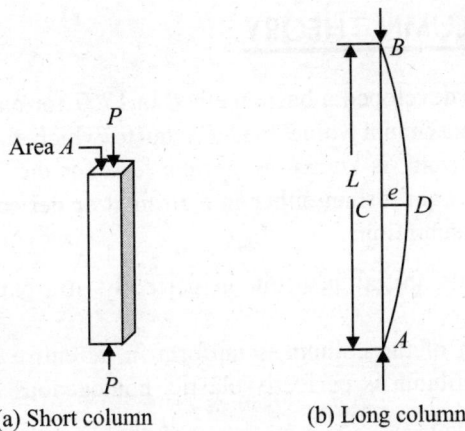

(a) Short column          (b) Long column

**Figure 17.1** Failure of short and long columns.

Consider the short column in Figure 17.1(a). Due to the load $P$, compressive load $P$, compressive load developed is:

$$P = \frac{P}{A}$$

Here $A$ is the cross-sectional area of the column. If this compressive force $P$ increased gradually, this short column will fail by crushing at a certain stage. The stress at which it fails by crushing is called *crushing stress* and $P$ at that stage is called *crushing load*. Thus short column fails by crushing. If $P_c$ is the crushing load, $\sigma_c$ is the crushing stress, then

$$\sigma_c = \frac{P_c}{A} \tag{17.1}$$

On the other hand, long column, shown in Figure 17.1(b), does not fail by crushing stress alone but also by bending or buckling stress. The long column buckles at certain load and this load is called *buckling load* or *crippling load*. This load is smaller than crushing load for long column. In Figure 17.1(b), $L$ is the length of the column, $P$ is the compressive load at which the column just buckles, $A$ is the cross-sectional area and $e$ is the maximum bending at centre of the column. Now $\sigma_0$ is the stress due to direct load and $\sigma_b$ is the stress due to bending at the centre. If $Z$ is the section modulus about the axis of bending, then:

$$\text{Maximum stress in the mid-section} = \sigma_0 + \sigma_b = \frac{P}{A} + \frac{Pe}{Z} \tag{17.2}$$

$$\text{And minimum stress} = \sigma_0 - \sigma_b = \frac{P}{A} - \frac{Pe}{Z} \tag{17.3}$$

Direct compressive stress $\sigma_0$ is negligible in case of a long column. Thus very long column is subjected to buckling stress only.

## 17.4 EULER'S COLUMN THEORY

It was Leonard Euler who developed a basic method in 1757 for analyzing long column based on the concept that the maximum value load $P_{cr}$ up to which the long column remains in position of neutral equilibrium, is known as *critical load* for the column and at critical load $P_{cr}$, column can remain in equilibrium either in a straight or deflected position. In his theory, he made the following assumptions:

(i) The column in its initial position is perfectly straight and the load is applied axially.
(ii) The cross section of the column is uniform in its entire length.
(iii) Material of the column is perfectly elastic, homogenous and isotropic and it obeys Hooke's law.
(iv) The length of the column is very large compared to its lateral dimensions.
(v) Direct stress is very small compared to the bending stress.
(vi) Column fails by buckling or crippling alone.
(vii) The self-weight of the column is negligible.

## 17.5 END CONDITIONS FOR LONG COLUMN AND SIGN CONVENTIONS

The important end conditions of the column are:

(i) Both ends of the column are hinged
(ii) One end is fixed and other end is free
(iii) Both ends are fixed
(iv) One end is fixed and other end is hinged or pinned.

Sign conventions of bending moment of column are explained in Figure 17.2(a) and in Figure 17.2(b).

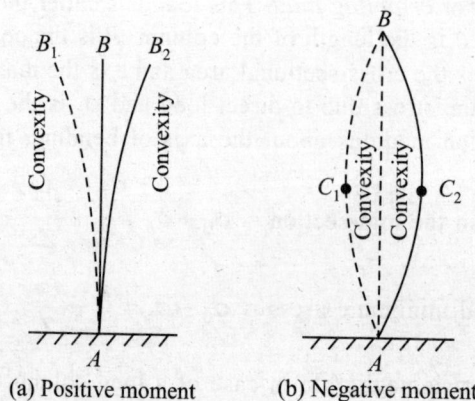

(a) Positive moment    (b) Negative moment

**Figure 17.2**   Sign coventions of bending moment.

A bending moment that bends the column so as to present the convexity towards the initial position $AB$ as shown in Figure 17.2(a) by $AB_1$ and $AB_2$, is regarded as positive moment.

A bending moment that bends the column so as to present the concavity towards the initial position $AB$ as shown in Figure 17.2(b) by $AC_1B$ and $A_2B$, is regarded as negative moment.

## 17.6 EQUATION OF CRIPPLING LOAD WHEN BOTH ENDS ARE HINGED

Figure 17.3 shows a column $AB$ whose ends are hinged at $A$ and $B$. Cross section of the column is uniform. The length of the column is $L$. The load $P$ is the crippling load at which the column just buckles in the curve form of $ACB$.

Consider a section at a distance $x$ from $A$. At $x$, column deflects a distance $y$ as shown in Figure 17.3.

The moment at this section due to crippling load

$$P = - P \cdot y \qquad (17.4)$$

(−ve sign as per sign convention discussed in Section 17.5) If $E$ is the Young's modulus and $I$ is the moment of inertia of the column section,

$$\text{Moment} = EI \frac{d^2y}{dx^2} \qquad (17.5)$$

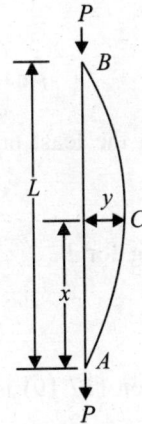

**Figure 17.3** Column $AB$ with both ends hinged.

Equating the moment from Eq. (17.4) and Eq. (17.5)

$$EI \frac{d^2y}{dx^2} = - Py$$

or

$$EI \frac{d^2y}{dx^2} + Py = 0 \qquad (17.6)$$

General solution of the differential Eq. (17.6) gives

$$y = C_1 \cdot \cos \left( x \sqrt{\frac{P}{EI}} \right) + C_2 \cdot \sin \left( x \sqrt{\frac{P}{EI}} \right) \qquad (17.7)$$

The constants of integration $C_1$ are $C_2$ obtained by putting the boundary conditions as:
At $A$, $x = 0$ and $y \cdot 0$
Substituting these two values in Eq. (17.7),

$$0 = C_1 \cdot \cos 0° + C_2 \sin 0°$$

or

$$0 = C_1 \times 1 + 0$$

$$\therefore \quad C_1 = 0$$

Equation (17.7) becomes

$$y = C_2 \cdot \sin \left( x \sqrt{\frac{P}{EI}} \right) \qquad (17.8)$$

At $B$, $x = L$ and $y = 0$

Substituting these boundary conditions in Eq. (17.8),

$$0 = C_2 \cdot \sin\left(L\sqrt{\frac{P}{EI}}\right) \tag{17.9}$$

In Eq. (17.9), $C_2$ cannot be zero because if $C_2$ is zero, $y$ in all section will be zero, i.e. column will not bend at all which is not true.

Hence

$$\sin\left(L\sqrt{\frac{P}{EI}}\right) = 0 = \sin 0 \quad \text{or} \quad \sin \cdot \pi \quad \text{or} \quad \sin \cdot 2\pi \dots$$

Taking the least practical values,

$$L\sqrt{\frac{P}{EI}} = \pi$$

Solving for $P$

$$P = \frac{\pi^2 EI}{L^2} \tag{17.10}$$

Equation (17.10) is the equation of crippling load of both hinged column.

## 17.7 EQUATION OF CRIPPLING LOAD WHEN ONE END FIXED AND OTHER FREE

A column $AB$ fixed in the end $A$ and free at $B$ is shown to deflect a distance $d$ under the action of crippling load $P$ is shown in Figure 17.4. Length of the column is $L$. Considering the deflection $y$ at a distance $x$ from $A$, moment of the crippling load at this section is:

$$\text{Moment} = P \cdot (d - y) \tag{17.11}$$

Moment in this situation is positive with reference to Figure 17.2.

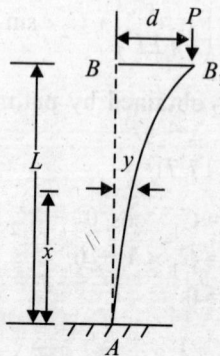

**Figure 17.4** Column $AB$ with one end fixed other end free.

Equating the moment in Eq. (17.11) with the moment in Eq. (17.5),

$$EI \frac{d^2y}{dx^2} = P(d - y)$$

or

$$EI \frac{d^2y}{dx^2} + P \cdot y = P \cdot D$$

or

$$\frac{d^2y}{dx^2} + \frac{Py}{EI} = \frac{Pd}{EI} \tag{17.12}$$

Writing

$$\left(\frac{P}{EI}\right)^{\frac{1}{2}} = m$$

Equation (17.12) may be written as: $\dfrac{d^2y}{dx^2} + m^2y = m^2D$

Solution of this differential equation can be obtained as:

$$y = C_1 \cdot \cos m \cdot x + C_2 \cdot \sin m \cdot x + d \tag{17.13}$$

Differentiating the above solution with respect to $x$,

$$\frac{dy}{dx} = C_1 m \sin mx + C_2 \cos m \cdot x \tag{17.14}$$

Boundary conditions at $A$ are: $x = 0$, $y = 0$ and the slope $\dfrac{dy}{dx} = 0$

Equation (17.14) now becomes,

$$0 = C_1 m \times 0 + C_2 \times 1$$

Hence

$$C_2 = 0$$

Again substituting the boundary conditions when $x = 0$, $y = 0$ using Eq. (17.13)

$$0 = C_1 \cdot \cos 0° + 0 + d = C_1 + d$$

∴
$$C_1 = -d$$

Again at $B$, $x = L$, $y = d$,
Equation (17.13) becomes

$$d = -d \cdot \cos m \cdot L + d$$

or
$$d \cdot \cos m \cdot L = 0$$

The deflection $d$ cannot be zero, hence $\cos m \cdot L = 0$

∴
$$mL = \frac{\pi}{2} \quad \text{or} \quad \frac{3\pi}{2} \dots$$

Taking the practical value,

$$mL = \left(\frac{P}{EI}\right)^{\frac{1}{2}} L = \frac{\pi}{2}$$

Crippling load

$$P = \frac{\pi^2 EI}{4L^2} \tag{17.15}$$

## 17.8   EQUATION OF CRIPPLING LOAD WHEN BOTH ENDS FIXED

Figure 17.5 shows a column $AB$ whose ends are fixed at $A$ and $B$.
Cross section of the column is uniform. The length of the column
is $L$. The deflected shape shows that slopes at both ends are zero.
Fixed-end moments are shown at both ends. The moment at section
$x$ is:

$$\text{Moment} = P \cdot y + M \tag{17.16}$$

Equating the moment in Eq. (17.16) with the moment in Eq. (17.5),

$$EI \frac{d^2 y}{dx^2} = -P \cdot y + M$$

or

$$\frac{d^2 y}{dx^2} + \frac{Py}{EI} = \frac{M}{EI} = \frac{M}{EI} \times \frac{P}{P} = \frac{P}{EI} \cdot \frac{M}{P}$$

Writing

$$\left(\frac{P}{EI}\right)^{\frac{1}{2}} = m$$

**Figure 17.5**  Column $AB$ with both ends fixed.

We have now

$$\frac{d^2 y}{dx^2} + m^2 y = m^2 \frac{M}{P}$$

Solution of this differential equation is:

$$y = C_1 \cdot \cos m \cdot x + C_2 \cdot \sin m \cdot x + \frac{M}{P} \tag{17.17}$$

Differentiating the above solution with respect to $x$,

$$\frac{dy}{dx} = -C_1 m \sin mx + C_2 \cos m \cdot x \tag{17.18}$$

Boundary conditions at $A$ are: $x = 0$, $y = 0$ and the slope $\frac{dy}{dx} = 0$
Equation (17.18) now becomes,

$$0 = -C_1 m \times 0 + C_2 \times 1$$

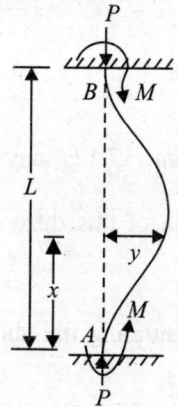

Hence
$$C_2 = 0$$
Again substituting the boundary conditions when $x = 0$, $y = 0$ in Eq. (17.17)

$$0 = C_1 \cdot \cos 0° + 0 + \frac{M}{P} = C_1 + \frac{M}{P}$$

$\therefore$
$$C_1 = -\frac{M}{P}$$

Substituting the values of $C_1 = -\dfrac{M}{P}$ and $C_2 = 0$ in Eq. (17.17)

$$y = -\frac{M}{P} \cdot \cos m \cdot x + 0 + \frac{M}{P}$$

or
$$y = -\frac{M}{P} \cdot \cos m \cdot x + \frac{M}{P} \qquad (17.19)$$

At the end $B$ of the column, $x = L$ and $y = 0$
Substituting these conditions in Eq. (17.19),

$$0 = -\frac{M}{P} \cdot \cos m \cdot L + 0 + \frac{M}{P}$$

or
$$\cos m \cdot L = \frac{M}{P} \times \frac{P}{M} = 1$$

$\therefore$
$$m \cdot L = 0,\ 2\pi,\ 4\pi,\ 6\pi \ ...$$

Taking the least practical value, $L\sqrt{\dfrac{P}{EI}} = 2\pi$

$\therefore$
$$P = \frac{4\pi^2 EI}{L^2} \qquad (17.20)$$

Equation (17.20) gives the crippling load equation of both ends fixed column.

## 17.9 EQUATION OF CRIPPLING LOAD WHEN ONE END FIXED AND OTHER HINGED

Figure 17.6 shows a column $AB$ whose one end is fixed at $A$ and other end at $B$ is hinged. Cross section of the column is uniform. The length of the column is $L$. The fixing moment $M$ at $A$ will give rise to horizontal reactions $H$ at each end $A$ and $B$ as shown in Figure 17.6 such that $H = M/L$.

From the deflected shape-shown in Figure 17.6, it can be written like the previous cases of crippling load as:

$$EI \frac{d^2 y}{dx^2} = -P \cdot y + H(L - x) \qquad (17.21)$$

or
$$\frac{d^2y}{dx^2} + \frac{P}{EI} \cdot y = \frac{P}{EI}\frac{H}{P}(L-x) \qquad (17.22)$$

Writing $\left(\dfrac{P}{EI}\right)^{\frac{1}{2}} = m$ like in the previous cases of Section 17.9

Equation (17.22) can be written as: $\dfrac{d^2y}{dx^2} + m^2 y = m^2\dfrac{H}{P}(L-x)$

Solution of above differential equation is:

$$y = C_1 \cdot \cos m \cdot x + C_2 \cdot \sin m \cdot x + \frac{H}{P}(L-x) \qquad (17.23)$$

At the fixed end, when $x = 0$, $y = 0$ and substituting this values in Eq. (17.23), we get

$$0 = C_1 + 0 + \frac{H}{P}L$$

Hence

$$C_1 = -\frac{HL}{P} \qquad (17.24)$$

Equation (17.23) becomes

$$y = -\frac{HL}{P}\cos mx + C_2 \cdot \sin m \cdot x + \frac{H}{P}(L-x) \qquad (17.25)$$

Differentiating (17.25) with respect to $x$,

$$\frac{dy}{dx} = -\frac{HL}{P}(-m\sin mx) + C_2 \cdot m\cos m \cdot x - \frac{H}{P} \qquad (17.26)$$

At $x = 0$, $\dfrac{dy}{dx} = 0$ as the point $A$ is fixed.

Equation (17.26) becomes:

$$0 = C_2 \cdot m \times 1 - \frac{H}{P}$$

Hence

$$C_2 = \cdot\frac{Hm}{P} \qquad (17.27)$$

Substituting the value of $C_1$ and $C_2$ from Eqs. (17.24) and (17.27) in Eq. (17.23),

$$y = -\frac{PL}{H}\cdot\cos m \cdot x + \frac{Hm}{P}\cdot\sin m \cdot x + \frac{H}{P}(L-x) \qquad (17.28)$$

At $B$, $x = L$ and $y = 0$, substituting these values in Eq. (17.28),

$$0 = -\frac{PL}{H}\cdot\cos m \cdot L + \frac{Hm}{P}\cdot\sin m \cdot L + \frac{H}{P}(L-L)$$

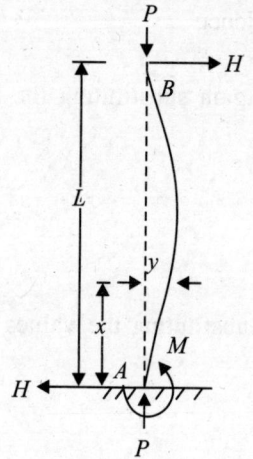

**Figure 17.6** Column $AB$ with one end fixed and other hinged.

or
$$\frac{PL}{H} \cdot \cos m \cdot L = \frac{Hm}{P} \cdot \sin m \cdot L$$

or
$$\tan mL = \frac{PL}{H}\frac{Hm}{P} = mL, \text{ i.e. } \tan \theta = \theta, \text{ solution of this equation is:}$$

$$mL = 4.5 \text{ radians}$$

or
$$L\left(\frac{P}{EI}\right)^{\frac{1}{2}} = 4.5$$

Squaring
$$L^2 \left(\frac{P}{EI}\right) = 20.25$$

But
$$2\pi_2 \approx 20.25$$

Hence
$$P = \frac{2\pi^2 EI}{L^2} \tag{17.29}$$

Equation (17.29) is the crippling load on column with one end fixed and other end hinged.

## 17.10    EFFECTIVE LENGTH OF COLUMN

It has been observed from the above four end conditions of column, Euler's crippling loads $P$ (or $P_{cr}$) have basically the same form. It can be expressed by all the Eqs. (17.10), (17.15), (17.20) and (17.13) as:

$$P_{cr} = \frac{\pi^2 EI}{L_e^2} \tag{17.30}$$

where $L_e$ is called the equivalent or effective length of column
It is seen that $L_e = L$, if both ends are hinged
$L_e = 2L$ if one end is fixed and other end is free
$L_e = \dfrac{L}{2}$, if both ends are fixed

$L_e = \dfrac{L}{\sqrt{2}}$, if one end is fixed and other end is hinged

## 17.11    SLENDERNESS RATIO OF COLUMN

The general critical Euler's crippling load is thus represented by Eq. (17.30), i.e.

$$P_{cr} = \frac{\pi^2 EI}{L_e^2}$$

Moment of inertia $I$ in the above equation can be replaced by in terms of area of cross-section $A$ and radius of gyration $k$ as $I = Ak^2$. Substituting this in Eq. (17.30)

$$P_{cr} = \frac{\pi^2 E A k^2}{L_e^2}$$

or

$$\frac{P_{cr}}{A} = \frac{\pi^2 E}{\left(\dfrac{L_e}{k}\right)^2}$$

The term $\dfrac{P_{cr}}{A}$ represents critical average stress $\sigma_{cr}$ or $p_{cr}$ in the member.

or

$$\sigma_{cr} \text{ or } p_{cr} = \frac{\pi^2 E}{\left(\dfrac{L_e}{k}\right)^2} \tag{17.31}$$

The term $\dfrac{L_e}{k}$ is called *slenderness ratio*.

Thus slenderness ratio $= \dfrac{\text{Length}}{\text{Radius of gyration}}$ (17.32)

Equation (17.31) shows that critical compress stress in a column is inversely proportional to the square of slenderness ratio. A plot of $\sigma_{cr}$ or $p_{cr}$ against $\dfrac{L_e}{k}$ is shown in Figure (17.7).

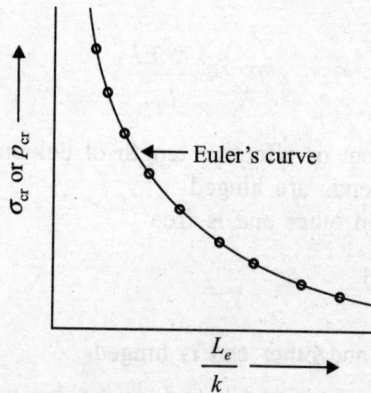

**Figure 17.7**   Plot of critical stress against slenderness ratio, i.e. Euler's curve.

It is called Euler's curve and the curve is asymptotic to both the axes. When $\dfrac{L_e}{k}$ tends to zero, $\sigma_{cr}$ or $p_{cr}$ tends to infinity.

## 17.12    LIMITATIONS OF EULER'S FORMULA

Consider Eq. (17.31) $\sigma_{cr}$ or $p_{cr} = \dfrac{\pi^2 E}{\left(\dfrac{L_e}{k}\right)^2}$ and Euler's curve in Figure 17.7.

In both, the equation and the curve show that if slenderness ratio is high, crippling stress will be higher. In any case crippling stress $\sigma_{cr}$ or $p_{cr}$ cannot be more than crushing strength of the column material. Thus if slenderness ratio falls to certain limit, Euler's formula gives a value of crippling stress which is more than the crushing stress. Thus there is a limit of slenderness ratio in order that crippling stress cannot be more then the crushing strength. The Euler's curve indicates that a low value of slenderness ratio causes increase in crippling stress. Thus in Euler's formula, there lies a limit of slenderness ratio $\dfrac{L}{k}$.

**EXAMPLE 17.1**    Calculate the safe compressive load on a hollow column having both ends fixed at 150 mm external diameter and 100 mm internal diameter. Length of the column is 6 m. Factor of safety is 3 and $E = 8 \times 10^4$ N/mm$^2$.

**Solution**    Data given are:

External diameter $D = 150$ mm, Internal diameter $d = 100$ mm
Length $L = 6\ m = 6000$ mm, $E = 8 \times 10^4$ N/mm$^2$, Factor of safety = 3

Now moment of inertia

$$I = \frac{\pi}{64}(D^4 - d^4) = \frac{\pi}{64}(150^4 - 100^4) = 19941750.24\ \text{mm}^4$$

Since both ends fixed, crippling load

$$P = \frac{4\pi^2 EI}{L^2} = \frac{4\pi^2 \times 8 \times 10^4 \times 19941750.24}{6000^2}$$

or                                $P = 1749486.097$ N

Using the factor of safety, safe compressive load

$$= \frac{1749486.097}{3}\ 583162.0324\ \text{N}$$

$$= 1749.486\ \text{kN}\quad\text{(Answer)}$$

**EXAMPLE 17.2**    Using the Euler's formula, calculate the critical stresses of columns having slenderness ratio 80 and 120 under the following conditions:

(i) Both ends hinged
(ii) Both ends fixed

**Solution**

(i) When both ends hinged, Euler's equation of crippling load is given by Eq. (17.10)

$$P = \frac{\pi^2 EI}{L^2}$$

In terms of effective length $L_e$, $L_e = L$

Slenderness ratio $= \dfrac{L_e}{k}$

Critical stress is $\sigma_{cr}$ or $p_{cr} = \dfrac{\pi^2 E}{\left(\dfrac{L_e}{k}\right)^2} = \dfrac{\pi^2 E}{80^2} = \dfrac{\pi^2 E}{6400}$     (Answer)

(ii) When both ends fixed, Euler's equation of crippling load is given by Eq. (17.20)

$$\therefore \qquad P = \frac{4\pi^2 EI}{L^2}$$

In terms of effective length $L_e$, $L_e = \dfrac{L}{2}$

Critical stress is $\sigma_{cr}$ or $p_{cr} = \dfrac{\pi^2 E}{\left(\dfrac{L_e}{k}\right)^2} = \dfrac{\pi^2 E}{\left(\dfrac{2L_e}{k}\right)^2} = \dfrac{\pi^2 E}{4\left(\dfrac{L_e}{k}\right)^2} = \dfrac{\pi^2 E}{4 \times (120)^2} = \dfrac{\pi^2 E}{57600}$

(Answer)

**EXAMPLE 17.3**  Determine the slenderness ratio for a steel column of solid circular section of diameter 150 mm and 3.5 m length.

**Solution**  Moment of inertia of the column $= I = \dfrac{\pi}{64} D^4 = \left(\dfrac{\pi}{64} \times 150^4\right)$ mm$^4$

Area of the column

$$= A = \frac{\pi}{4} D^2 = \frac{\pi}{4} \times 150^2 \text{ mm}^2$$

If $k$ is the radius of gyration, then, $I = Ak^2$

or

$$\frac{\pi}{64} \times 150^4 = \frac{\pi}{4} \times 150^2 \times k^2$$

$$\therefore \qquad k = 37.5 \text{ mm}$$

Slenderness ratio $= \dfrac{L}{k} = \dfrac{3.5 \times 1000}{37.5} = 93.333$     (Answer)

**EXAMPLE 17.4**  A column of timber section 150 mm × 200 mm is 6 m long, both ends being fixed. Find the safe load for the column. Use Euler's formula and allow a factor of safety of 3. Take $E = 17500$ N/mm$^2$.

***Solution***   Area of column

$$A = 150 \times 200 = 3000 \text{ mm}^2$$

Length of the column

$$L = 6 \ m = 6000 \text{ mm}$$
$$b = 150 \text{ mm}$$
$$d = 200 \text{ mm}$$
$$E = 17500 \text{ N/mm}^2.$$

Moment of inertia

$$I_{xx} = \frac{1}{12} bd^3 = \frac{1}{12} \times 150 \times 200^3 = 10000 \times 10^4 \text{ mm}^4$$

Moment of inertia

$$I_{yy} = \frac{1}{12} db^3 = \frac{1}{12} \times 200 \times 150^3 = 5625 \times 10^4 \text{ mm}^4$$

Since $I_{yy}$ is less than $I_{xx}$, column will buckle in $y$-direction and value of $I_{yy}$ will be taken. Since both ends fixed,

$$P = \frac{4\pi^2 EI}{L^2} = \frac{4\pi^2 \times 17500 \times 5625 \times 10^4}{6000^2} \text{ N}$$

$$P = 1079487.981 \text{ N}$$

Using a factor of safety of 3,

Safe load on the column $= \dfrac{1079487.981}{3} = 359829.3271 \text{ N} = 359.829 \text{ kN}$    (Answer)

***EXAMPLE 17.5***   A steel bar of solid circular cross section is 50 mm diameter. The bar is pinned at each end and subjected to an axial compression. If the length of the bar is 2 m and $E = 2 \times 10^{11}$ N/m², determine the buckling load using Euler's equation.

***Solution***   Diameter of column

$$D = 50 \text{ mm}$$

Length of the column

$$L = 2 \ m = 2000 \text{ mm}$$
$$E = 2 \times 10^{11} \text{ N/m}^2 = 2 \times 10^5 \text{ N/mm}^2$$

Moment of inertia

$$I = \frac{\pi}{64} D^4 = \frac{\pi}{64} \times 50^4 = 306796.1596 \text{ mm}^4$$

Column is pinned or hinged at both ends and hence Euler's equation of buckling load is

$$P = \frac{\pi^2 EI}{L^2} = \frac{\pi^2 \times 2 \times 10^5 \times 306796.1596}{2000^2} = 151397.8354 \text{ N} = 151.3978 \text{ kN}$$    (Answer)

**EXAMPLE 17.6**   Determine the slenderness ratio for a timber column of 200 mm × 250 mm in cross section and 8 m long.

**Solution**   Size of timber column

$$b = 200 \text{ mm}, \ d = 250 \text{ mm}$$

Length of the column

$$L = 8, \ m = 8000 \text{ mm}$$

Area of the column

$$A = 200 \times 250 = 50000 \text{ mm}^2$$

Moment of inertia

$$I_{xx} = \frac{1}{12} bd^3 = \frac{1}{12} \times 200 \times 250^3 = 260416666.7 \text{ mm}^4 = 260.41667 \text{ m}^4$$

Moment of inertia

$$I_{yy} = \frac{1}{12} db^3 = \frac{1}{12} \times 250 \times 200^3 = 166666666.7 \text{ mm}^4 = 166.667 \text{ m}^4$$

Since $I_{yy}$ is less than $I_{xx}$, column will buckle in $y$-direction and value of $I_{yy}$ will be taken. If $k$ is the radius of gyration, then, $I = Ak^2$
Then putting the values of moment of inertia $I_{yy}$ and area $A$,

$$166666666.7 = 50000 \cdot k^2$$

Hence

$$k = 57.735 \text{ mm} \qquad \text{(Answer)}$$

**EXAMPLE 17.7**   A hollow mild steel tube 50 mm internal and 6 mm thick is used as a column with both ends hinged. The length of tube is 6.2 m; find the safe load on it with a factor of safety of 3. Take $E = 2 \times 10^5$ N/mm$^2$.

**Solution**

$$\text{Internal diameter } d = 50 \text{ mm}$$
$$\text{Thickness } t = 6 \text{ mm}$$

∴

$$\text{External diameter } D = d + 2t = (50 + 2 \times 6) = 62 \text{ mm}$$
$$\text{Length of the column} = 6.2 \text{ m} = 6200 \text{ mm}$$
$$\text{Young's modulus } E = 2 \times 10^5 \text{ N/mm}^2$$
$$\text{Factor of safety (FOS)} = 3$$

Moment of inertia

$$I = \frac{\pi}{64} (D^4 - d^4) = \frac{\pi}{64} (62^4 - 50^4) = 418535.5397 \text{ mm}^4$$

Crippling load for both ends hinged

$$P = \frac{\pi^2 EI}{L^2} = \frac{\pi \times 2 \times 10^5 \times 418535.5397}{6200^2} = 6841.145 \text{ N}$$

$$\text{Safe load on the column} = \frac{P}{\text{FOS}} = \frac{6841.145}{3} = 2280.383 \text{ N} \qquad \text{(Answer)}$$

***EXAMPLE 17.8*** A circular bar is used as both ends pin-jointed column. The length of the column is 5 m. The same bar when freely supported gives a midspan deflection of 10 mm under a load of 80 N at centre. Find the Euler's critical load.

***Solution*** Data given are:

$$\text{Length } L = 5 \text{ m} = 5000 \text{ mm}$$
$$\text{Weight at midspan } W = 80 \text{ N}$$
$$\text{Deflection } d = 10 \text{ mm}$$

Deflection of freely supported bar at midspan due load $W$ is $d = \dfrac{WL^2}{48EI}$

$$\therefore \qquad EL = \frac{WL^3}{48d} = \frac{80 \times 5000^3}{48 \times 10} = 2.083333 \times 10^{10} \text{ N/mm}^2$$

Euler's critical load with both ends pin-jointed

$$P = \frac{\pi^2 EI}{L^2} = \frac{\pi^2 \times 2.083333 \times 10^{10}}{5000^2}$$

$$P = 8224.67 \text{ N} \qquad \text{(Answer)}$$

## 17.13   CONCLUSION

At the beginning of the chapter, presentations are noted on the causes of failure of short and long columns under load by crushing and buckling respectively. Euler's column theory, his assumptions in the development of the theory and end-conditions for long column and sign conventions are discussed. Euler's equations of buckling load on column under four different end-conditions are derived step by step in a very easy-to-follow methods of solution of differential equations with end boundary conditions. Limitations of Euler's theory, effective length of long column, slenderness ratio, plot critical stress against slenderness ratio showing the Euler's curve are presented. Quite a few numerical examples are solved to show the use of Euler's equations. Some chapter-end questions and numerical problems are given with answers at the end for practice.

## EXERCISES

**17.1** State the assumptions of Euler's column theory. How the failure of a shot and of a long column takes place?

**17.2** What do you understand by effective or equivalent length of column? Give the effective length of column under four different end-conditions?

**17.3** What is the slenderness ratio of column? Draw a rough sketch of Euler's curve. What are the limitations of Euler's formula?

**17.4** Prove that critical crippling stress by Euler's formula if given by $\sigma_{cr}$ or $p_{cr} = \dfrac{\pi^2 E}{\left(\dfrac{L_e}{k}\right)^2}$.

**17.5** Derive the equation for crippling load when both ends are fixed.

**17.6** Derive the expression for critical load for a long column with both ends hinged loaded by an axial compressive force at each end.

**17.7** A column of 60 m diameter is 3 m long. One end of the column is fixed while other end is hinged. If Young's modulus determine safe compressive load by Euler's formula allowing a factor of safety 3.5.                                      (Answer    79740 N)

**17.8** A steel tube of 30 mm internal diameter and 4 mm thickness is used as column of length 8 m. If $E = 2.1 \times 10^5$ N/mm$^2$, find collapsing load by Euler's formula.

(Answer    8110 N)

**17.9** A pin-jointed on both ends steel column has cross section of 60 mm × 100 mm. If $E = 2 \times 10^5$ N/mm$^2$, and critical stress is $E = 250$ N/mm$^2$, find shortest length $L$ of steel column.                                      (Answer    1539.058 mm)

# Index